Methods in Cell Biology

Cell Polarity and Morphogenesis

Volume 139

Series Editors

Leslie Wilson
Department of Molecular, Cellular and Developmental Biology
University of California
Santa Barbara, California

Phong Tran
University of Pennsylvania
Philadelphia, USA &
Institut Curie, Paris, France

Methods in Cell Biology

Cell Polarity and Morphogenesis

Volume 139

Edited by

Thomas Lecuit
IBDM (Institut de Biologie du Développement de Marseille)
UMR 7288, Aix-Marseille Université & CNRS, Marseille, France;
Collège de France, Paris, France

ELSEVIER

ACADEMIC PRESS
An imprint of Elsevier
elsevier.com

Academic Press is an imprint of Elsevier
50 Hampshire Street, 5th Floor, Cambridge, MA 02139, United States
525 B Street, Suite 1800, San Diego, CA 92101-4495, United States
125 London Wall, London EC2Y 5AS, United Kingdom
The Boulevard, Langford Lane, Kidlington, Oxford OX5 1GB, United Kingdom

First edition 2017

ISBN: 978-0-12-809373-3
ISSN: 0091-679X

Publisher: Zoe Kruze
Acquisition Editor: Zoe Kruze
Editorial Project Manager: Colford Hannah
Production Project Manager: Surya Narayanan Jayachandran
Designer: Miles Hitchen

Contents

Contributors

B. Aigouy
IBDM (Institut de Biologie du Développement de Marseille) UMR 7288, Aix-Marseille Université & CNRS, Marseille, France

N. Bertaux
Institut Fresnel, Aix Marseille Université, Centrale Marseille, CNRS, Marseille, France

R. Buđa
Ruđer Bošković Institute, Zagreb, Croatia

C. Cadart
Institut Curie/Institut Pierre-Gilles de Gennes, Paris, France

C. Collinet
IBDM (Institut de Biologie du Développement de Marseille) UMR 7288, Aix-Marseille Université & CNRS, Marseille, France

A. Das
Johns Hopkins University, Baltimore, MD, United States

S. De Renzis
European Molecular Biology Laboratory (EMBL), Heidelberg, Germany

K. Guevorkian
Université de Strasbourg, Centre National de la Recherche Scientifique, UMR 7104, Institut National de la Santé et de la Recherche Médicale, U964, Illkirch, France

G. Guglielmi
European Molecular Biology Laboratory (EMBL), Heidelberg, Germany

O. Hamant
Laboratoire de Reproduction et Développement des Plantes, ENS-Lyon, INRA, CNRS, UCBL, Université de Lyon, 46 allée d'Italie, 69364 Lyon Cedex 07, France

Y. Hamon
Centre d'Immunologie de Marseille Luminy, Aix Marseille Université, Inserm, CNRS, Marseille, France

H.-T. He
Centre d'Immunologie de Marseille Luminy, Aix Marseille Université, Inserm, CNRS, Marseille, France

M. Kaksonen
University of Geneva, Geneva, Switzerland

M. Le Berre
Institut Curie/Institut Pierre-Gilles de Gennes, Paris, France

R. Li
Johns Hopkins University, Baltimore, MD, United States

J.-L. Maître
Institut Curie, PSL Research University, CNRS, UMR 3215, INSERM, U934, Paris, France

S. Mailfert
Centre d'Immunologie de Marseille Luminy, Aix Marseille Université, Inserm, CNRS, Marseille, France

D. Marguet
Centre d'Immunologie de Marseille Luminy, Aix Marseille Université, Inserm, CNRS, Marseille, France

M. Merkel
Syracuse University, Syracuse, NY, United States

N. Minc
Institut Jacques Monod, CNRS, UMR 7592, Paris, France

S. Monnier
Institut Curie/Institut Pierre-Gilles de Gennes, Paris, France

A. Picco
University of Geneva, Geneva, Switzerland

M. Piel
Institut Curie/Institut Pierre-Gilles de Gennes, Paris, France

V. Racine
QuantaCell, Pessac, France

M. Rauzi
Université Côte d'Azur, CNRS, INSERM, iBV, Nice, France

A. Sagner
The Francis Crick Institute, London, United Kingdom

P. Singh
Johns Hopkins University, Baltimore, MD, United States

T. Stanislas
Laboratoire de Reproduction et Développement des Plantes, ENS-Lyon, INRA, CNRS, UCBL, Université de Lyon, 46 allée d'Italie, 69364 Lyon Cedex 07, France

H. Tanimoto
Institut Jacques Monod, CNRS, UMR 7592, Paris, France

O. Thouvenin
Institut Langevin/ESPCI Paris Tech, Paris, France

I.M. Tolić
Ruđer Bošković Institute, Zagreb, Croatia

J. Traas
Laboratoire de Reproduction et Développement des Plantes, ENS-Lyon, INRA, CNRS, UCBL, Université de Lyon, 46 allée d'Italie, 69364 Lyon Cedex 07, France

L. Venkova
Institut Curie/Institut Pierre-Gilles de Gennes, Paris, France

K. Vukušić
Ruđer Bošković Institute, Zagreb, Croatia

E. Zlotek-Zlotkiewicz
Institut Curie/Institut Pierre-Gilles de Gennes, Paris, France

CHAPTER

1

A user's guide for characterizing plasma membrane subdomains in living cells by spot variation fluorescence correlation spectroscopy

S. Mailfert*, Y. Hamon*, N. Bertaux§, H.-T. He*, D. Marguet*[,1]

**Centre d'Immunologie de Marseille Luminy, Aix Marseille Université, Inserm, CNRS, Marseille, France*

§Institut Fresnel, Aix Marseille Université, Centrale Marseille, CNRS, Marseille, France

[1]Corresponding author: E-mail: marguet@ciml.univ-mrs.fr

CHAPTER OUTLINE

Methods in Cell Biology, Volume 139, ISSN 0091-679X, http://dx.doi.org/10.1016/bs.mcb.2016.12.002

Abstract

Due to the intrinsic molecular Brownian agitation within plasma membrane and the vast diversity of membrane components, it is expected that the plasma membrane organization is highly heterogeneous with the formation of local complex multicomponent assemblies of lipids and proteins on different time scales. Still, deciphering this lateral organization on living cells and on the appropriate length and temporal scales has been challenging but is crucial to advance our knowledge on the biological function of the plasma membrane. Among the methodological developments based on biophotonics, the spot variation FCS (svFCS), a fluorescent correlation spectroscopy (FCS)—based method, has allowed the significant progress in the characterization of cell membrane lateral organization at the suboptical level, including to providing compelling evidence for the in vivo existence of lipid-dependent nanodomains. The aim of this chapter is to serve as a guide for setting and applying the svFCS methodology to study the plasma membrane of both adherent and nonadherent cell types.

INTRODUCTION

The fluid mosaic model provides a central paradigm still relevant to depict a general and consensual scheme of the membrane organization (Singer & Nicolson, 1972). It emphasizes two major features characterizing the cell membranes, i.e., the asymmetric organization and lateral mobility within the membrane bilayer. In that respect, experiments evidencing the lateral diffusion of membrane components (Frye & Edidin, 1970) have been seminal to elaborate this model.

At first, we can consider that Brownian molecular agitation acts by creating a tendency at homogenizing cell membrane. However, the diversity of lipids and proteins present in cell membranes counteracts this tendency for homogeneity by constantly achieving selective molecular interactions and cooperative effects among the membrane components. Thus, local inhomogeneities are expected but their characterization in living cells has challenged this field of research. Among the major observations, there have been studies pointing out the broad spatiotemporal scales over which the local inhomogeneities occur within the plasma membrane. In addition, the role of such local organizations on the regulation of biological functions has been subject of numerous investigations.

From a methodological point of view, this field of research has benefitted from the development of fluorescence-based approaches. Among them, the fluorescence correlation spectroscopy (FCS) introduced in the 70th (see for review Elson, 2013) has been of general interest by providing the sensitivity, noninvasiveness and robustness required to perform observations at the plasma membrane of living cells (Chiantia, Ries, & Schwille, 2009; Schwille, Haupts, Maiti, & Webb, 1999). The general principle of FCS is based on recording the temporal fluorescence fluctuations within a confocal volume of illumination characterized by the beam waist of illumination. The interpretation of these fluctuations yields information on the dynamics of the molecular system under scrutiny such as diffusion, interaction or chemical reaction [see for details on FCS theory, experimental applications and methods of analysis (Tetin, 2013a, 2013b)].

Conventional FCS observations, i.e., at a single waist of observation, can discriminate subsets of molecular components when their respective diffusion coefficients D differ by a factor below ~ 1.6 (Meseth, Wohland, Rigler, & Vogel, 1999). We overcome this limitation by varying the volume of illumination to analyze the diffusion time as a function of the area intercepting the cell membrane. This approach, namely, the spot variation FCS (svFCS) is reminiscent from the fluorescence recovery after photobleaching (FRAP) analysis made at different length scales (Yechiel & Edidin, 1987). svFCS has provided particular insight to identify possible molecular confinement occurring at the plasma membrane (He & Marguet, 2011; Lenne et al., 2006). Changing the spot of illumination can be achieved by underfilling the back aperture of the objective (Wawrezinieck, Lenne, Marguet, & Rigneault, 2004) or by the z-scan method (Masuda, Ushida, & Okamoto, 2005). Finally, it is now possible scaling the FCS observations below the optical diffraction limits by the use of single nanometric aperture technology (Wenger et al., 2007) or by shaping the volume with STED nanoscopy (Eggeling et al., 2009; Honigmann et al., 2014).

Theoretical models have been developed to predict how geometrical constraints such as the presence of adjacent or isolated domains affect the svFCS observations (Ruprecht, Wieser, Marguet, & Schutz, 2011; Wawrezinieck, Rigneault, Marguet, & Lenne, 2005). These investigations have allowed significant progress in the characterization of cell membrane lateral organization at the suboptical level for different kind of receptors (Blouin et al., 2016; Chakrabandhu et al., 2007; Guia et al., 2011) and have provided compelling evidence for the in vivo existence of lipid-dependent nanodomains (Lasserre et al., 2008; Lenne et al., 2006; Wenger et al., 2007).

Here, we detail a step-by-step protocol for performing svFCS observations at the membrane of living cells.

1. BUILDING A SENSITIVE AND ROBUST HOMEMADE SPOT VARIATION FLUORESCENCE CORRELATION SPECTROSCOPY/FLUORESCENCE RECOVERY AFTER PHOTOBLEACHING SETUP

Performing svFCS and FRAP measurements does not require highly sophisticated optical setups. We present here with details a simple setup that enables the observation on living cells with sufficient sensitivity (see Fig. 1 and Table 1). In the following paragraphs, *z* refers to the optical axis, *xy* to the perpendicular axes, waist size to the radius of the point spread function (PSF).

1.1 DESCRIPTION OF THE OPTICAL BENCH AND SIGNAL COLLECTION

1. The power of a 488 nm CW laser cooled by an air-cooling system (Fan) is controlled by an acousto-optic modulator (AOM).
2. The two mirrors (M) periscope ensures a proper coupling into an optical fiber (OF) through a high precision fiber coupler (C). The beam height is adjusted according to the microscope entry.
3. An achromatic doublet collimates the beam.
4. A motorized iris (MI) adjusts the beam diameter filling the back aperture of the objective to tune the effective numerical aperture that consequently modulates the resulting waist size (Wawrezinieck et al., 2004). With a 40 × /1.2NA water-immersion objective, the waist size is adjustable between 200 and 400 nm.
5. A dichroic mirror (DM) reflects the excitation beam to the microscope entry. A high quality and thick (≈5 mm) DM designed for laser applications limits the PSF deformations.
6. The laser power at the microscope entry is adjusted with a power meter (P) on the light reflected by a mirror (M) on a motorized flipper mount (MF) controlled by the acquisition board (ACQ) to switch between the power measurement and sample illumination.
7. The excitation beam is directed to the objective lens (O) through a 100% microscope port. Minimizing undesired alterations on the excitation beam and optimizing the collection of light are obtained by avoiding the presence of tube lenses within the beam path.
8. The focused laser beam is parked at the appropriate location on the biological sample with a *xyz* piezoelectric nanopositioning stage.
9. A 50 μm pinhole (PH) is placed at the Fourier plane of a first lens (AD_2, f = 150 mm) to remove out-of-focus signal. The signal is focused on the single photon counting module (SPCM) with a second relay lens (AD_3, f = 35 mm).
10. A fluorescence filter (F) rejects inappropriate light without introducing substantial light path deviation to the detector.
11. The signal is recorded by a multitau hardware correlator and an ACQ. Data are further analyzed (see Section 4 for detailed procedures).

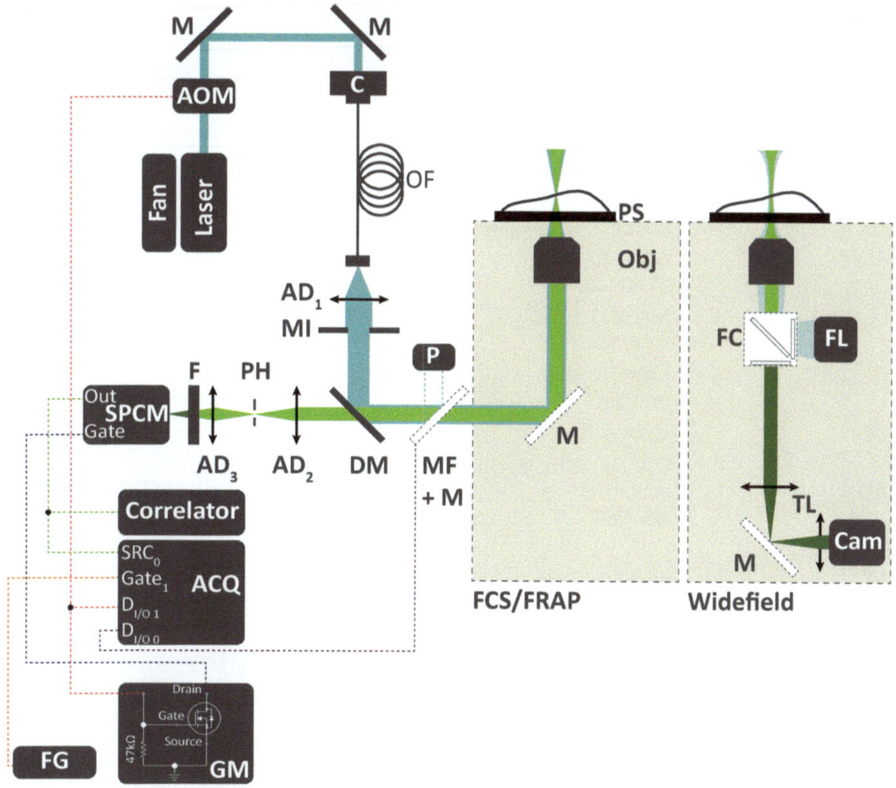

FIGURE 1 Optical setup for spot variation fluorescence correlation spectroscopy and fluorescence recovery after photobleaching (FRAP).

The power of a 488 nm laser beam is tuned by an acousto-optic modulator (AOM). The output beam from the optical fiber is collimated and sized by a motorized iris, before being focused on the sample by the microscope objective via a dichroic mirror. The backward fluorescent light from the sample is collected through a confocal light path and recorded on a single photon counting module (SPCM) after emission wavelength selection (F). The signal is autocorrelated by a multitau hardware correlator. The acquisition board (ACQ) ensures the control and synchronization of the different devices: SPCM, AOM, homemade gating module (GM), an external clock generated by the function generator (FG) or the mirror flipper (MF). For the FCS or FRAP measurements, the tube lenses are absent keeping the beam collimation and ensuring better photon collection. Widefield observations of the biological samples illuminated with a mercury fluorescence lamp are recorded with a sCMOS camera. See Section 1 and Table 1 for further details.

Table 1 Supplies for Homemade Fluorescence Correlation Spectroscopy—Fluorescence Recovery After Photobleaching System

Name	Description	Company	Reference
Laser	Laser 488 nm, CW	Coherent Inc.	Sapphire 488 LP 100 mW
Fan	Air-cooling fan	N/A	N/A
FG	Function generator	GW	GFG-8019G
AOM	Acousto-optic modulator	IntraAction Corp.	ME
M	Mirror, 100% reflection	Thorlabs Inc.	PF10-03-P01
C	Laser to fiber coupler	OZ Optics	HPUC-23AF-400/ 700-P-6AC-2
OF	Optical fiber	OZ Optics	QPMJ-3AF3S-488-3.5/125-3-3-1
AD_{1-3}	Achromatic doublet	Thorlabs Inc.	AD_1: AC254-060-A AD_2: AC508-150-A AD_3: AC254-035-A
MI	Motorized iris	Standa	8SMC4-USB-B8-1
DM	Dichroic mirror	Chroma Technology Corp.	ZT488rdc
MF	Mirror flipper	Thorlabs Inc.	MFF101/M
P	Power meter	Thorlabs Inc.	PM100D + S120C
Obj	Objective lens	Zeiss	C-apochromat 40X/ 1.2 W Corr M27
PS	*xyz* piezoelectric stage	Physik Instrumente	E710.4CL + P-527.3CL
PH	Pinhole	Thorlabs Inc.	P50S
F	Fluorescence filter	Omega Optical	535AF45
SPCM	Single photon counting module	Perkin Elmer	SPCM-AQR13
Correlator	2-channels hardware correlator	Correlator.com	Flex02-12D
ACQ	Acquisition board	National Instruments	BNC2121 + NI PCI-6602
GM	Gating module with MOSFET	Homemade	2N7000 MOSFET
FL	Mercury fluorescence lamp	Zeiss	HBO
FC	Fluorescence cube	Zeiss	Filter set 13
Cam	sCOMS camera	Hamamatsu	Orca Flash 4.0 LT
TL	Tube lenses	Zeiss	N/A
MIC	Microscope	Zeiss	Axiovert 200M
T	Temperature chamber	Life Imaging Services	Cube & Box

1.2 COMMON PITFALLS AND PRACTICAL CONSIDERATIONS

The following points are critical in designing a robust and sensitive optical setup:

1. Simplify the optical path as much as possible.
2. Use high precision *xy* and *z* micrometric adjustment mounts or translation stages to warrant excitation light collimation and reduce loss of emitted photons. Best performances are obtained by using separated adjustment mounts for the SPCM, lenses and pinhole.
3. Install the whole microscope into an incubator box with a temperature controller (The Cube & Box, Life Imaging Services) to obtain the best temperature control.
4. Protect the detector (SPCM) from damaging high flux of photons by linking its gate to the ACQ through a gating module (GM) based on a MOSFET transistor. If the recorded signal is higher than the typical 500,000 kHz, the SPCM gate receives the signal to shut down the detector.
5. When performing the FRAP measurements, the SPCM receives the same signal to its gate during the photobleaching step and the AOM the opposite one to increase the laser power. The temporal resolution during the FRAP measurements is based on an external clock provided by a function generator (FG) (typically set at 10 kHz for an acquisition rate of 100 μs).

2. SAMPLE PREPARATIONS AND LABELING

Successful svFCS measurements depend on multiple parameters that need to be set before the fluorescence signal acquisition itself. Here, we discuss several general aspects regarding the sample preparation, fluorescence labeling, and pharmacological treatments.

2.1 GENERAL CONSIDERATIONS ON SAMPLE PREPARATIONS

svFCS measurements have been successfully conducted on different cell types, either adherent or not, primary or established cell lines. We normally perform the measurements on cells seeded in 8-well Lab-Tek chambers.

For adherent cells, the wells are precoated overnight in complete medium. Cells are seeded at 1.5–2.0 × 10^4 cells/well, 24 h prior to FCS acquisitions, in phenol red-free culture medium supplemented with fetal bovine serum. The culture reaches 30–50% confluence before the FCS experiments. Cells are washed two to three times in Hank's Balanced Salt Solution with 10 mM 4-(2-hydroxyethyl)-1-piperazineethanesulfonic acid (HEPES) pH 7.4 (HBSS/HEPES), and allowed to rest at least 30 min prior to recordings.

For nonadherent cells, they are seeded in wells after having performed an extemporal precoating with poly-L-lysine (0.01%) for 30 min at 37°C. Cells are sedimented by gravitation. Special cares should be taken with regards to the cell incubation conditions during measurements or adjusted to experimental situations (i.e., if no CO_2 source is available). For instance, using HBSS/HEPES instead of culture medium for measurements on T cells causes significant stress with cellular deformation or membrane movement. As a rule, those cells are serum starved overnight in DMEM F12 (Gibco) supplemented with Nutridoma SP (Sigma), which standardizes the experimental conditions compatible with further metabolic or pharmacologic treatments (see below).

2.2 USE OF FLUORESCENT PROTEINS

Fluorescent proteins (FP) are widely used as reporters in cell biology. Although frequently used in FCS studies, they are not devoid of drawbacks or limitations (Baens et al., 2006; Chudakov, Matz, Lukyanov, & Lukyanov, 2010). Monomeric forms of fluorescent proteins (Zacharias, 2002) should be favored although we did not find major aggregation effect at the concentrations compatible with FCS measurements (unpublished data). However, the length of the peptide linking the FP to the protein of interest might alter the diffusion measurements and the nature and strength of a membrane confinement (unpublished results). More importantly, it is mandatory to ascertain an absence of physiological alterations once the molecule of interest is tagged with an FP but also an absence of competition of the recombinant protein with the endogenous one. This problem can be circumvented by silencing the endogenous protein by RNA interference, natural variants, or genetically modified cells.

Transfection is performed $\approx$24 h prior svFCS measurements directly in the well for adherent cells. The expression kinetics of the protein needs to be taken into account. The amounts of DNA should be optimized to keep the expression of the FP-tagged protein compatible with svFCS measurements (usually $\approx$10 times lower than the supplier's recommendations). This also facilitates to come close to the physiological expression levels. Alternatively, establishing stable transfected cell lines overcomes the variability of transfection efficiencies. Then, it is preferable to select a stable subpopulation on the basis of its low fluorescence level using fluorescence-activated cell sorting (FACS) as well as to minimize selecting abnormal clones.

2.3 MEMBRANE LABELING WITH LIPID PROBES. IMPACT ON FLUORESCENCE CORRELATION SPECTROSCOPY MEASUREMENTS

Fluorescent lipid analogues (i.e., boron-dipyrromethene-tagged lipids) or lipophilic tracers (i.e., carbocyanine or aminostyryl dyes) can be used for svFCS. Nevertheless, cautions should be made when using standard bovine serum albumin (BSA)-exchange based protocols. Indeed, BSA acting as a nonspecific lipid carrier can alter the physicochemical composition of cell membranes. This effect is also potentially problematic when BSA is used to saturate nonspecific binding sites for antibodies.

2.4 SURFACE LABELING OF NONADHERENT CELLS WITH ANTIBODIES PRIOR TO FLUORESCENCE CORRELATION SPECTROSCOPY MEASUREMENTS

Possible direct observations on plasma membrane proteins are performed by labeling cells with well-characterized fluorescent antibodies. The genetically or biochemically engineered monoclonal and monovalent antibodies are very interesting tools because of their high specificity and low aggregation potential (Andrew, 2000).

The choice of the fluorophores depends on the laser sources but also of their own brightness, photostability, and amphiphilic properties (Hughes, Rawle, & Boxer, 2014). Alexa Fluor (Invitrogen) or ATTO (ATTO-TEC GmbH) series functionalized by reactive chemical residues (i.e., NHS ester, iodoacetamide) are easily coupled covalently on a large variety of biomolecules including antibody-derived proteins. Labeling conditions should be carefully adjusted (statistically one dye per molecule) to minimize a loss in binding affinity or specificity.

As an example, we detail here a protocol for cultured T lymphocytes.

1. Harvest T cells by mild centrifugation at 90 g for 8 min at room temperature (RT) and suspend cells at 10^6/mL in DMEM F12 without phenol red, 1% Nutridoma-SP (DMEM-Nutridoma) in presence of fluorescent Fabs.
2. Split cells into 200 μL/well of a 96-well round-bottom plate.
3. Incubate the plate for 30–60 min at 10°C which strongly reduces the cell endocytosis.
4. Centrifuge the plate at 90 g for 8 min at 10°C.
5. Remove the supernatant, wash the pellet once with DMEM-Nutridoma before being resuspended in 200 μL of DMEM-Nutridoma.
6. Store the plate above ice in the dark up to 4 h before measurements.
7. Seed the cells into a 0.01% poly-L-lysine coated well with DMEM-Nutridoma at RT up to 5 min prior to FCS measurements.

Note

- The fluorescent probe concentration is adjusted to label a number of molecules of interest compatible with FCS measurements i.e., less than a few tens of nanomolar concentration.

2.5 PHARMACOLOGICAL TREATMENTS

When characterizing the origin of molecular confinements in the membrane, we use pharmacological drugs perturbing the actin meshwork organization or lipid, especially sphingolipid and cholesterol, composition.

2.5.1 Latrunculin A treatment (actin meshwork disassembly)

For adherent cells (i.e., COS-7 cells)

1. Wash the cells in DMEM-Nutridoma in the well.
2. Incubate cells in 0.1 μM latrunculin A (from an intermediate 25× dilution in dimethyl sulfoxide (DMSO) of a stock solution at 2.5 mM stored at −20°C) for 30 min at 37°C.
3. Start the FCS acquisitions in presence of latrunculin A.

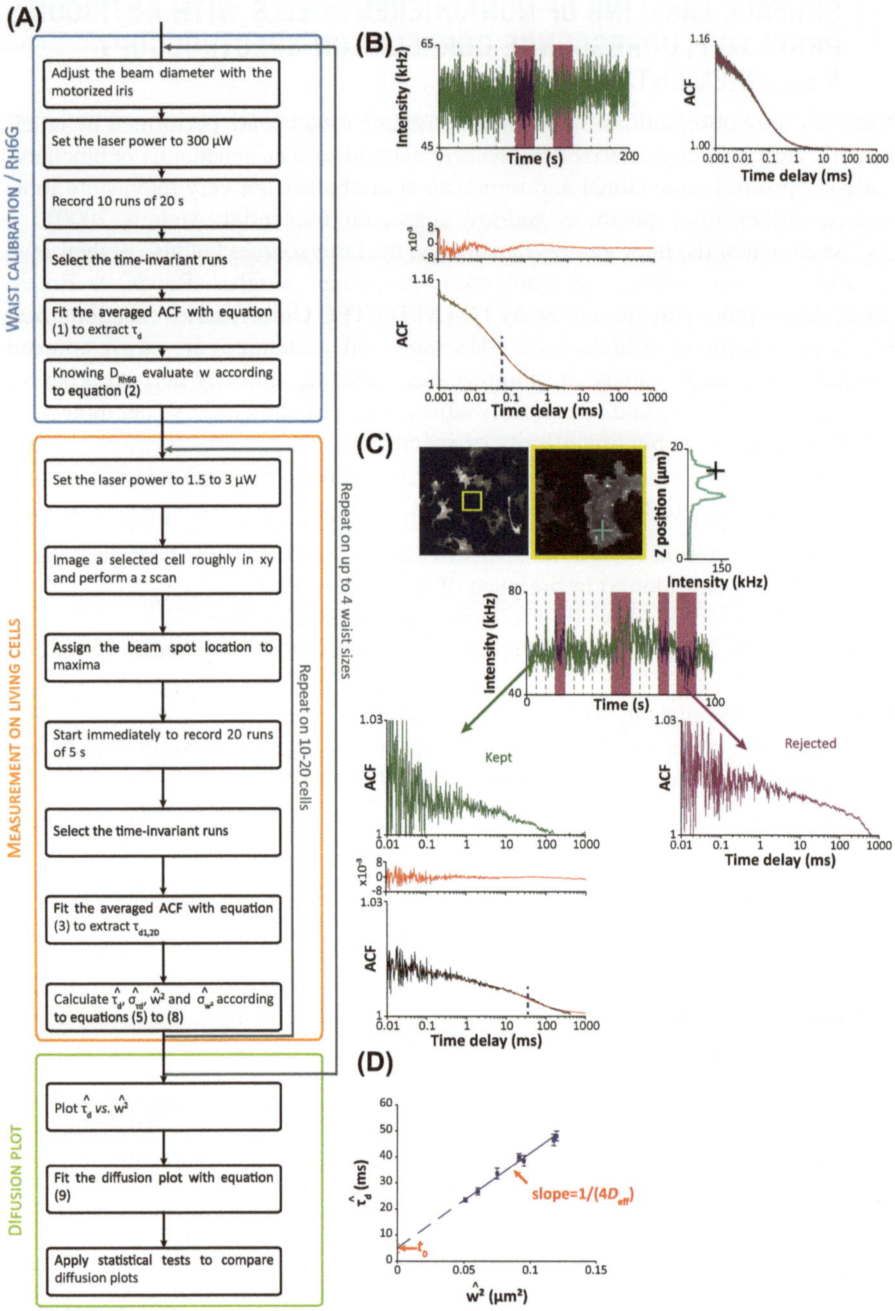

FIGURE 2 Spot variation fluorescence correlation spectroscopy (svFCS) recording and data analysis workflow.

(A) Schematic view of the svFCS data recording and analyzing workflow divided into three main steps: the waist size calibration, the recording of FCS measurements on cells, and the data analysis procedure (see Section 3 for the detailed procedures). (B) Examples of raw and

For nonadherent cells (i.e., T cells)

1. Resuspend cells in 200 μL DMEM-Nutridoma in the 96-well plate.
2. Mix with 5 μM latrunculin A diluted in 200 μL DMEM-Nutridoma.
3. Seed the cells into a 0.01% poly-L-lysine coated well for 30 min at 37°C.
4. Start the FCS acquisitions in presence of latrunculin A.

2.5.2 Sphingolipid and cholesterol metabolism inhibition

1. Resuspend cells (i.e., 2.5×10^6 T cells; 5.0×10^5 COS-7) in 30 mL complete culture medium with 25 μM myriocin (stock solution 20 mM in DMSO, stored at −80°C).
2. Seed cells into a 250 mL flask for 48 h.
3. If needed, electroporation or transfection shall be performed after 36 h but kept in complete culture medium with 25 μM myriocin for the remaining 12 h.
4. Cells are harvested and washed three times in DMEM F12.
5. Adjust the cell density at 0.8×10^6 cells/mL for T cells or 5×10^4 cells/mL for COS-7 in DMEM-Nutridoma in presence of 25 μM myriocin, 10 μM zaragozic acid (stock solution 5 mM in water, stored at −80°C).
6. Seed the cells in a flask for nonadherent cells, or directly in well for adherent cells.
7. Culture cells for 16 h.
8. Start the FCS acquisitions in presence of myriocin and zaragozic acid whose actions are reversible.

3. FLUORESCENCE CORRELATION SPECTROSCOPY RECORDING

Fig. 2A resumes the workflow to record the measurements. Steps described in Sections 3.1 and 3.2 have to be primarily realized before measurements on living cells.

analyzed data obtained for the calibration with standard solution of Rh6G. The selection of the autocorrelation function (ACF) is based on the scrutiny for the time-invariance of each individual trace (10 × 20 s). Once selected, the averaged resulting curve is fitted to compute the waist size knowing the diffusion time of the Rh6G. (C) For the cell measurements, the spot of illumination at reduced laser power is parked on the upper membrane by performing a coarse *xy* image of the fluorescence distribution, a *z* scan and finally a fine *xyz* positioning. Then, the acquisition procedure is similar to the one performed with the Rh6G solution. Precise *xyz* location of the area of interest leads to accurate measurements. The procedure is repeated for at least more than 10 cells/waist and four to five different waist sizes. Scale bars: 50 μm. (D) The diffusion plot represents the variation of the diffusion time τ_d as function of the square of the waist w^2 and fitted to extract the *y*-axis intercept (t_0) and the effective diffusion coefficient (D_{eff}).

3.1 PRELIMINARY RECORDING

1. Prewarm the microscope chamber at the desired temperature (i.e., at 37°C for mammalian cells).
2. If the chamber is not equipped with a CO_2 supply, buffer the culture media with 10 mM HEPES.

3.2 SPOT SIZE CALIBRATION

Fig. 2A and B resume the calibration procedure and illustrate the corresponding FCS recording and fitting, respectively.

1. Prepare a standard solution of rhodamine 6G at 20.8 mM (Rh6G, 10 mg/mL) and store aliquots at −20°C.
2. Drop 200 μL of 20.8 pM Rh6G obtained by cascade (1/10) dilutions on a glass cover−slip laid down directly on the pupil of a water-immersion objective.
3. Adjust the 488 nm laser beam power to 300 μW.
4. Record fluorescence fluctuations of Rh6G during 10 runs of 20 s.
5. The autocorrelation functions (ACFs) buildup for each individual runs are examined for stability (see below for the criteria).
6. An average ACF of the kept independent recordings is fitted with a 3-D diffusion model including triplet state, and calculate the average diffusion time τ_d.
7. Calculate the waist value w^2 knowing the diffusion coefficient of Rh6G ($D = 280\ \mu m^2/s$) in aqueous solution at 37°C and according to Eq. (1) (Magde, Elson, & Webb, 1974).

$$\tau_D = w^2/4D \tag{1}$$

8. Resume the procedure to determine the spot size following any tuning modification of the laser beam diameter.

3.3 FLUORESCENCE CORRELATION SPECTROSCOPY ACQUISITIONS AT THE PLASMA MEMBRANE OF LIVING CELLS

1. Adjust the 488 nm laser beam power between 1.5 and 3 μW (i.e., less than 2 kW/cm^2), as function of the cell type and fluorescent probe.
2. Equilibrate cells 10 min at 37°C before starting the measurements.
3. Identify cells under widefield fluorescence illumination for appropriate (usually low) signals.
4. Perform a rough *xy* confocal image of the selected cell.
5. Perform a *z* scan usually in the nuclear area to maximize the separation between the top and bottom plasma membrane.
6. Assign the confocal spot at one of the maximal fluorescence intensity and start as quick as possible the acquisitions (choosing the top or bottom plasma membrane depends on biological considerations).

7. Collect 20 consecutive runs of 5 s (or 10 runs of 10 s). These parameters are set empirically considering the biological characteristic of the molecule of interest (diffusion coefficient, fast recycling, or internalization) and the respective photophysical characteristics of the probe.
8. Confocal spot can only be repositioned once on the same cell.
9. Perform 10 to 15 series of recordings on about 10 different cells.
10. Data are fitted as described in Section 4.

Note

- Measurements do not last more than 30 min on the same sample.

3.4 SPOT FLUORESCENCE RECOVERY AFTER PHOTOBLEACHING MEASUREMENTS AT THE PLASMA MEMBRANE OF LIVING CELLS

It is important to determine the mobile fraction M_f of molecules contributing to the fluorescence fluctuations recorded on the time scale of an FCS measurement. This is done on the same setup (see Section 1) by performing conventional spot FRAP recording (Axelrod, Koppel, Schlessinger, Elson, & Webb, 1976) to characterize the mobile fraction M_f and diffusion coefficient D_{frap} of the molecule of interest. The procedure is similar to the one previously described for svFCS, except that the intensity of the laser power is modulated by the AOM during the sequence.

1. Same as described for svFCS acquisition at Section 3.3, points 1 to 6 except for the selection of cells with approximately 3–10 times higher fluorescent signal than for svFCS (cell expressing more fluorescent chimera or with antibodies coupled with a ratio protein/fluorophore $\geq$2).
2. Once the confocal spot is assigned at a proper position, start the FRAP recording with a prebleaching, bleaching pulse, and postbleaching phases. The sampling frequency, pulse duration and intensity have to be adjusted to the experimental model.
3. Repeat FRAP measurement no more than twice on the same cell.

Notes

- The same setup allows to perform FRAP measurements with same xy and z positioning accuracy and detection sensitivity as for svFCS measurements, but also a high temporal resolution.
- Regular FRAP parameters for T lymphocytes or COS-7 cells are of 1 s prebleaching recording at regular 1.5–3 μW laser illumination, 1 ms bleach pulse at 500 times higher power intensity, and 3–5 s of recording at initial laser illumination during the recovery phase.
- Sampling frequency at 100 μs is critical to provide accurate and robust fitting.
- The mobile fraction M_f and diffusion coefficient D_{frap} are computed from the fluorescence recovery curve fitting. Around 25 to 35 measurements get sustainable statistics.

- D_{frap} uses to be consistent with the one measured by FCS at the same waist.
- M_f is often higher than published values mainly due to the absence of fluorescence recording delay after the bleaching pulse, which increases the robustness of its evaluation.

4. DATA ANALYSIS AND INTERPRETATION

The FCS data acquisition and analyses are illustrated for waist calibration (Fig. 2B), a set of data for a fluorescently labeled membrane component (Fig. 2C) and the corresponding diffusion plot obtained by performing a set of acquisition at different spot size (Fig. 2D).

4.1 DATA ANALYSIS AND FITTING PROCEDURES

The diffusion plot linking the diffusion time τ_d to the square of the waist w^2 of the confocal spot is obtained by computing data for at least four different waists. For each waist, the averaged τ_d value results from fitting around 20 ACFs obtained on 10 different cells; each ACF representing the averaged records from selective scrutiny of the 20 consecutive runs of 5 s (see Section 3.3). This selection aims at discarding runs where major cell membrane movements, high fluorescent peaks due to aggregates or endocytic vesicles disturb the stationary photon counting rate. The y-axis intercept, namely t_0 and the inverse of the slope, the effective diffusion coefficient D_{eff} are computed by a linear fitting of the diffusion plot weighted by the standard deviations on both parameters (see below).

The analysis procedures are described below:

1. ACF of individual runs is selected on a time-invariant criterion of the trace or discarded if abrupt variation or slope is evidenced.

The fitting accuracy is obtained from data recorded on low fluorescently labeled samples as the ACF amplitude is inversely proportional to the number of molecules present within the observation volume.

2. The selected curves are averaged, and the resulting ACF fitted by a realistic model:
   a. Waist size determination—one species diffusing in 3-D with a triplet state component (Eq. 2).

$$G(\tau) = 1 + \left(1 + \frac{T}{1-T} e^{\left(-\frac{\tau}{t_T}\right)}\right) \times \frac{1}{N} \times \left(\frac{1}{1+\tau/\tau_{d,3D}}\right)\left(\frac{1}{\sqrt{1+s^2 \times \tau/\tau_{d,3D}}}\right) \tag{2}$$

where T is the fraction of nonfluorescent dyes in triplet state with t_T its characteristic time, N the average number of molecules within the observation volume, $\tau_{d,3D}$ the mean diffusion time, and s the structural parameter (ratio between the waist size in xy and z) usually fixed to 0.2, a value close to the theory and experimentally validated by measuring the PSF on beads.

b. Two species, 2-D diffusion model (Eq. 3), usually for cells expressing eGFP-tagged protein.

$$G(\tau) = 1 + \frac{1}{N}\left(\frac{A}{1+\tau/\tau_{d1,2D}} + \frac{1-A}{1+\tau/\tau_{d2,2D}}\right) \tag{3}$$

where N is the average number of molecules within the observation volume, $\tau_{d1,2D}$ and $\tau_{d2,2D}$ the mean diffusion time of species 1 and 2, and A their relative ratio. Usually, the second species is detectable due to broken link between the eGFP and the membrane protein and leads to a fast diffusion time in-between 0.5 to 1 ms.

c. One species, 2-D diffusion model (Eq. 4), usually for cells labeled with fluorescently conjugated Fabs.

$$G(\tau) = 1 + \frac{1}{N}\left(\frac{1}{1+\tau/\tau_{d,2D}}\right) \tag{4}$$

where N is the average number of molecules within the observation volume, $\tau_{d,2D}$ the mean diffusion time.

3. This single set of measurement is repeated on 10 to 20 different cells providing the same number of mean diffusion times.

a. The mean diffusion time $\widehat{\tau_d}$ weighted by the number of selected ACFs/run for a total of P runs, and the standard deviation $\sigma(\widehat{\tau_d})$ are given by Eqs. (5) and (6), respectively:

$$\widehat{\tau_d} = \frac{\sum_{i=1}^{P} \tau_d(i) \times Selected(i)}{\sum_{i=1}^{P} Selected(i)} \tag{5}$$

$$\sigma\left(\widehat{\tau_d}\right) = \sqrt{\frac{\sum_{i=1}^{n}\left(\widehat{\tau_d} - \tau_d(i)\right)^2 \times Selected(i)}{P \times \sum_{i=1}^{P} Selected(i)}} \tag{6}$$

where $\tau_d(i)$ is the diffusion time for the ith run and *Selected* (i) the number of kept ACFs.

4. The standard deviation $std(\widehat{w^2})$ on the waist value w is estimated from the consecutive measurements done with the rhodamine 6G by computing $\widehat{w^2}$ and $\sigma(\widehat{w^2})$ as follow:

$$\widehat{w^2} = \frac{\sum_{i=1}^{P} \tau_d(i) \times 4 \times D_{Rh6G} \times Selected(i)}{\sum_{i=1}^{P} Selected(i)} \tag{7}$$

and

$$\sigma\left(\widehat{w^2}\right) = \sqrt{\frac{\sum_{i=1}^{n}\left(\widehat{w^2} - \tau_d(i) \times 4 \times D_{Rh6G}\right)^2 \times Selected(i)}{P \times \sum_{i=1}^{P} Selected(i)}} \tag{8}$$

For each w, we determine the $\sigma(\widehat{w^2})$ according to the linear dependency of $\sigma(\widehat{w^2})$ on $\widehat{w^2}$.

5. The diffusion plot which represents $\widehat{\tau_d}$ versus $\widehat{w^2}$ with their respective standard deviations, is fitted with a linear relationship taking into account the standard deviations in both axes [for more details, see (Billaudeau et al., 2013)]. The extrapolated intercept on the y-axis, namely t_0, and the inverse of the slope, i.e., the effective diffusion coefficient D_{eff} are obtained from a linear fit of the diffusion plot given in Eq. 9 (Wawrezinieck et al., 2005):

$$\tau_d = t_0 + \left(\frac{1}{4 \times D_{\text{eff}}}\right) \times w^2 \tag{9}$$

Note

- All of the homemade procedures developed on IGOR Pro (WaveMetrics Inc.) or MATLAB (Mathworks) are available on request.
- As the temporal resolution of the hardware correlator is 80 MHz, the amount of data collected cannot be easily saved as raw data. We only work on the computed ACF provided by the correlator and a nontemporally resolved trace.
- Discarding data obtained at large waists (i.e., $w > 350$ nm) is generally more pronounced due to concentration effect and more frequent experimental perturbations.
- For the experiments with fluorescently conjugated Fabs, the fitting Eq. (3) can be also used instead of Eq. (4) to identify the fraction of dissociated probe from the membrane.

4.2 DATA INTERPRETATION

The key issue when reporting the svFCS observables, e.g., the t_0 and D_{eff} values, for a molecule of interest localized at the plasma membrane, is to identify the origin of the deviations as regards a strict Brownian behavior. Initially, we found experimentally and confirmed by simulations of confined diffusion that the t_0 is positive if the membrane components are trapped in isolated domains but negative when they are trapped in a network of barriers (Lenne et al., 2006; Saxton, 2005; Wawrezinieck et al., 2005). This is resumed in Fig. 3A.

To characterize the molecular mechanism underlying such deviations, svFCS analyses are performed under conditions disturbing the plasma membrane

(A)

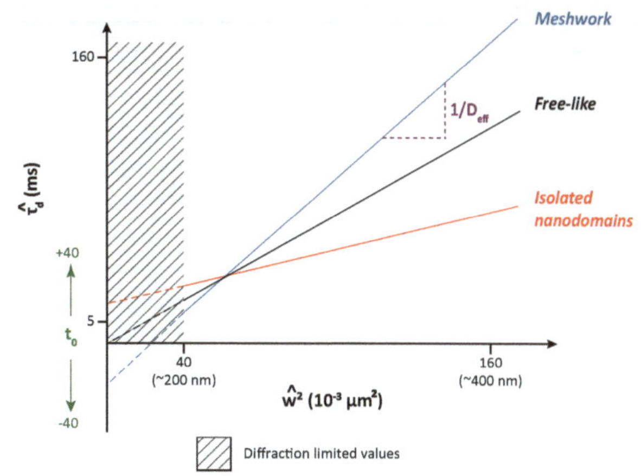

(B)

#	Experimental conditions	$t_0 \pm$ SD (ms)	$D_{eff} \pm$ SD (µm²/s)
1	Non treated	11.8 ± 2.5	1.04 ± 0.13
2	Jasplakinolide	11.7 ± 0.7	1.03 ± 0.10
3	COase 1 U	-0.2 ± 1.8	0.74 ± 0.04

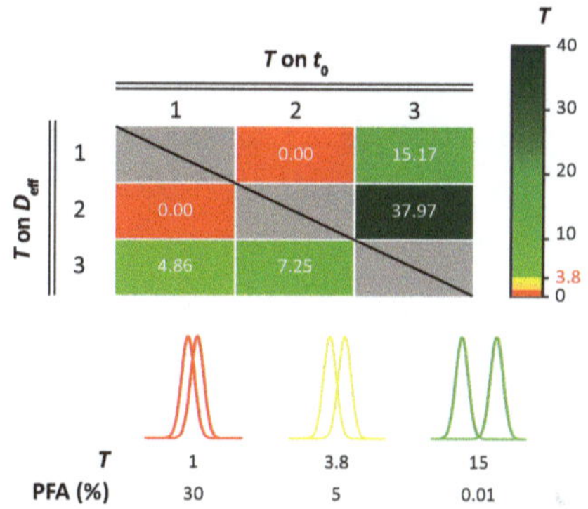

FIGURE 3 Diffusion models for membrane organization.

(A) The diffusion plots for the different models of molecular diffusion at the plasma membrane. The free-like diffusion model shows a straight linearity of τ_d with w^2 and a null intercept with the time axis ($t_0 = 0$). In the presence of isolated domains where molecules are transiently trapped in domains, the extrapolated t_0 intercept on the time axis becomes positive. With adjacent domains, the presence of permeable barriers prevents molecules to freely diffuse, and the extrapolated t_0 becomes negative. See for details (He & Marguet, 2011; Lenne et al., 2006; Wawrezinieck et al., 2005). (B) t_0 and D_{eff} for the eGFP-GPI protein in COS-7 under different experimental conditions (Lenne et al., 2006) (left panel). T values calculated for each pairwise comparison (upper right panel). Illustration of the overlaps for two normal distributions of same variances observed at different T values; the corresponding probability of false alarm (PFA) which is computed with an incomplete gamma function delimited by a specific T_0 threshold is indicated below (lower right panel).

organization. For instance, this can be done by applying pharmacological treatments destabilizing the actin-based cytoskeleton meshwork (Kusumi et al., 2005) or by using drugs altering the metabolism of specific lipids (Blouin et al., 2016; Lasserre et al., 2008; Lenne et al., 2006).

4.3 COMPARISON OF SPOT VARIATION FLUORESCENCE CORRELATION SPECTROSCOPY RESULTS

We also need to establish if the t_0 or D_{eff} values reported under two experimental conditions are different. Thus, we have developed a test discriminating if the data originate from the same statistical process or not by opposing the H_2 hypothesis (the two values are different) to the H_1 hypothesis (the two values are not different) at a threshold set above a probability of false alarm (PFA).

Let consider a general case with two independent measurements m_A and m_B with σ_A^2 and σ_B^2 their respective variance and assuming a Gaussian distribution.

Under the H_1 hypothesis, the Gaussian distributions of m_A and m_B have same mean μ_0 but at the precision of their respective variances: $m_A \sim \mathcal{N}(\mu_0, \sigma_A^2)$ and $m_B \sim \mathcal{N}(\mu_0, \sigma_B^2)$. The log-likelihood of the H_1 hypothesis is written as:

$$\text{logPDF}(H_1) = -\frac{1}{2}\log(2\pi\sigma_A\sigma_B) - \frac{1}{2\sigma_A^2}(m_A - \mu_0)^2 - \frac{1}{2\sigma_B^2}(m_B - \mu_0)^2 \quad (10)$$

However, under H_2 hypothesis, m_A and m_B follow a Gaussian distribution at their respective means μ_A^0 and μ_B^0 and at the precision of their respective variances: $m_A \sim \mathcal{N}(\mu_A^0, \sigma_A^2)$ and $m_B \sim \mathcal{N}(\mu_B^0, \sigma_B^2)$. Thus, the log-likelihood of the H_2 hypothesis is written as:

$$\text{logPDF}(H_2) = -\frac{1}{2}\log(2\pi\sigma_A\sigma_B) - \frac{1}{2\sigma_A^2}(m_A - \mu_A^0)^2 - \frac{1}{2\sigma_B^2}(m_B - \mu_B^0)^2 \quad (11)$$

Because there is only one experimental mean value per estimated parameter, m_A and m_B are the trivial solution of the maximum likelihood estimations of μ_A^0 and μ_B^0, respectively and according to Eq. (11), the corresponding log-likelihood logPDF(H_2) as:

$$\text{logPDF}(H_2) = -\frac{1}{2}\log(2\pi\sigma_A\sigma_B) \quad (12)$$

Under H_1, the estimation of μ_0 is obtained by minimizing the equation:

$$\frac{\partial(\text{logPDF}(H_1))}{\partial(\mu_0)} = 0 \quad (13)$$

Thus

$$\widehat{\mu_0} = m_A \frac{\sigma_B^2}{\sigma_A^2 + \sigma_B^2} + m_B \frac{\sigma_A^2}{\sigma_A^2 + \sigma_B^2} \quad (14)$$

Accordingly, the log-likelihood logPDF(H_1) is defined as follow:

$$\text{logPDF}(H_1) = -\frac{1}{2}\log(2\pi\sigma_A\sigma_B) - \frac{1}{2\sigma_A^2}\left(\widehat{\mu_0} - m_A\right)^2 - \frac{1}{2\sigma_B^2}\left(\widehat{\mu_0} - m_B\right)^2 \quad (15)$$

By inserting in Eq. (15) $\widehat{\mu_0}$ as defined in Eq. (14) and by defining $\Delta m = m_A - m_B$, we obtain:

$$\text{logPDF}(H_1) = -\frac{1}{2}\log(2\pi\sigma_A\sigma_B) - \frac{1}{2}\frac{\Delta m^2}{\sigma_A^2 + \sigma_B^2} \qquad (16)$$

Thus, the generalized likelihood ratio test (GLRT) aims to ascertain that the two values are different by calculating a contrast value T, the difference between $\text{logPDF}(H_2)$ and $\text{logPDF}(H_1)$:

$$T = 2(\text{logPDF}(H_2) - \text{logPDF}(H_1)) = \frac{\Delta m^2}{\sigma_A^2 + \sigma_B^2} \qquad (17)$$

Under H_1 hypothesis, Δm is zero-mean and because m_A and m_B are following a Gaussian distribution, $\Delta m \sim \mathcal{N}(0, \sigma_A^2 + \sigma_B^2)$. It is easy to obtain a standard normal distribution by dividing Δm by $\sqrt{\sigma_A^2 + \sigma_B^2}$. The product corresponds to $\sqrt{T}$, which thus follows a standard normal distribution $\mathcal{N}(0, 1)$ and so, T follows a χ^2 at one degree of freedom.

Thus, the probability $P(T > T_0|H_1)$ is evaluated for a specified T_0 threshold expressed as a PFA. The more the contrast value T is high, the more the compared values are different from each other.

This is illustrated on experimental data (Fig. 3B). On the left, the t_0 or D_{eff} values for the eGFP-tagged onto the glycophosphatidylinositol-anchoring signal originating from the decay accelerating factor (eGFP-GPI) in COS-7 are recapitulated under three different experimental conditions (Lenne et al., 2006). The right table resumes the T values returned by the test for pairwise comparison of t_0 and D_{eff}. We can conclude for instance, that T reported for the comparison of t_0 and D_{eff} measured under the experimental conditions #1 and #2 does not allow to consider the data as different; the H_2 hypothesis is not verified and the stabilization of the actin-based cytoskeleton by jasplakinolide has no significant effect on the membrane confinement of the eGFP-GPI protein. On the contrary, the results observed in conditions #1 and #3 are significantly different with a PFA suggesting that, for eGFP-GPI, the nanodomain organization generated by dynamic partitioning process is no longer existing at the plasma membrane following the cholesterol oxidase treatment.

Note

- The test runs under MATLAB and is freely available for academic research on demand.

CONCLUSION

Because FCS displays single-molecule sensitivity and high temporal resolution, it has been applied to analyze molecular diffusion at the plasma membrane of living cells under noninvasive conditions. Performing FCS measurements at variable spot sizes, e.g., svFCS, allows characterizing two pertinent observables, t_0 and D_{eff}, whose variations are particularly informative to discriminate the geometry of membrane confinements, i.e., isolated versus contiguous domains. Combining such methodological approach with pharmacological treatments has contributed to demonstrate unambiguously the existence of dynamic lipid-dependent nanodomains at the plasma membrane of very different cell types. Finally, our studies based on svFCS observations performed since 2005 emphasize that proteins and lipids operate in concert within the dynamic membrane organization for controlling transmembrane receptor signaling in living cells.

ACKNOWLEDGMENTS

We thank our colleagues at the CIML and Institut Fresnel for their constant support. This work is supported by grants from the Ministère de l'Enseignement Supérieur et de la Recherche (ANR-10-BLAN-1214, ANR-10-INBS-04 France BioImaging and ANR-11-LABX-0054 Labex INFORM), the Fondation pour la Recherche Médicale (FRM-DEQ-20090515412), Aix-Marseille Université (ANR-11-IDEX-0001-02 A*MIDEX), and institutional funding from the Centre National de la Recherche Scientifique, the Institut National de la Santé et de la Recherche Médicale.

REFERENCES

Andrew, S. M. (2000). Enzymatic digestion of monoclonal antibodies. *Methods in Molecular Medicine, 40*, 325–331. http://dx.doi.org/10.1007/978-1-59259-076-6_25.

Axelrod, D., Koppel, D. E., Schlessinger, J., Elson, E., & Webb, W. W. (1976). Mobility measurement by analysis of fluorescence photobleaching recovery kinetics. *Biophysical Journal, 16*(9), 1055–1069. http://dx.doi.org/10.1016/S0006-3495(76)85755-4.

Baens, M., Noels, H., Broeckx, V., Hagens, S., Fevery, S., Billiau, A. D., … Marynen, P. (2006). The dark side of EGFP: defective polyubiquitination. *PLoS One, 1*, e54. http://dx.doi.org/10.1371/journal.pone.0000054.

Billaudeau, C., Mailfert, S., Trombik, T., Bertaux, N., Rouger, V., Hamon, Y., … Marguet, D. (2013). Probing the plasma membrane organization in living cells by spot variation fluorescence correlation spectroscopy. *Methods in Enzymology, 519*, 277–302. http://dx.doi.org/10.1016/B978-0-12-405539-1.00010-5.

Blouin, C. M., Hamon, Y., Gonnord, P., Boularan, C., Kagan, J., Viaris de Lesegno, C., … Lamaze, C. (2016). Glycosylation-dependent IFN-γR partitioning in lipid and actin nanodomains is critical for JAK activation. *Cell, 166*(4), 920–934. http://dx.doi.org/10.1016/j.cell.2016.07.003.

Chakrabandhu, K., Herincs, Z., Huault, S., Dost, B., Peng, L., Conchonaud, F., … Hueber, A. O. (2007). Palmitoylation is required for efficient Fas cell death signaling. *EMBO Journal, 26*(1), 209–220. http://dx.doi.org/10.1038/sj.emboj.7601456.

Chiantia, S., Ries, J., & Schwille, P. (2009). Fluorescence correlation spectroscopy in membrane structure elucidation. *Biochimica et Biophysica Acta, 1788*(1), 225–233. http://dx.doi.org/10.1016/j.bbamem.2008.08.013.

Chudakov, D. M., Matz, M. V., Lukyanov, S., & Lukyanov, K. A. (2010). Fluorescent proteins and their applications in imaging living cells and tissues. *Physiological Reviews, 90*(3), 1103−1163. http://dx.doi.org/10.1152/physrev.00038.2009.

Eggeling, C., Ringemann, C., Medda, R., Schwarzmann, G., Sandhoff, K., Polyakova, S., ... Hell, S. W. (2009). Direct observation of the nanoscale dynamics of membrane lipids in a living cell. *Nature, 457*(7233), 1159−1162. http://dx.doi.org/10.1038/nature07596.

Elson, E. L. (2013). Brief introduction to fluorescence correlation spectroscopy. *Methods in Enzymology, 518*, 11−41. http://dx.doi.org/10.1016/B978-0-12-388422-0.00002-9.

Frye, L. D., & Edidin, M. (1970). The rapid intermixing of cell surface antigens after formation of mouse-human heterokaryons. *Journal of Cell Science, 7*(2), 319−335.

Guia, S., Jaeger, B. N., Piatek, S., Mailfert, S., Trombik, T., Fenis, A., ... Ugolini, S. (2011). Confinement of activating receptors at the plasma membrane controls natural killer cell tolerance. *Science Signaling, 4*(167), ra21. http://dx.doi.org/10.1126/scisignal.2001608.

He, H. T., & Marguet, D. (2011). Detecting nanodomains in living cell membrane by fluorescence correlation spectroscopy. *Annual Review of Physical Chemistry, 62*, 417−436. http://dx.doi.org/10.1146/annurev-physchem-032210-103402.

Honigmann, A., Mueller, V., Ta, H., Schoenle, A., Sezgin, E., Hell, S. W., & Eggeling, C. (2014). Scanning STED-FCS reveals spatiotemporal heterogeneity of lipid interaction in the plasma membrane of living cells. *Nature Communications, 5*, 5412. http://dx.doi.org/10.1038/ncomms6412.

Hughes, L. D., Rawle, R. J., & Boxer, S. G. (2014). Choose your label wisely: water-soluble fluorophores often interact with lipid bilayers. *PLoS One, 9*(2), e87649. http://dx.doi.org/10.1371/journal.pone.0087649.

Kusumi, A., Nakada, C., Ritchie, K., Murase, K., Suzuki, K., Murakoshi, H., ... Fujiwara, T. (2005). Paradigm shift of the plasma membrane concept from the two-dimensional continuum fluid to the partitioned fluid: high-speed single-molecule tracking of membrane molecules. *Annual Review of Biophysics and Biomolecular Structure, 34*, 351−378. http://dx.doi.org/10.1146/annurev.biophys.34.040204.144637.

Lasserre, R., Guo, X. J., Conchonaud, F., Hamon, Y., Hawchar, O., Bernard, A. M., ... He, H. T. (2008). Raft nanodomains contribute to Akt/PKB plasma membrane recruitment and activation. *Nature Chemical Biology, 4*(9), 538−547. http://dx.doi.org/10.1038/nchembio.103.

Lenne, P. F., Wawrezinieck, L., Conchonaud, F., Wurtz, O., Boned, A., Guo, X. J., ... Marguet, D. (2006). Dynamic molecular confinement in the plasma membrane by microdomains and the cytoskeleton meshwork. *EMBO Journal, 25*(14), 3245−3256. http://dx.doi.org/10.1038/sj.emboj.7601214.

Magde, D., Elson, E. L., & Webb, W. W. (1974). Fluorescence correlation spectroscopy. II. An experimental realization. *Biopolymers, 13*(1), 29−61. http://dx.doi.org/10.1002/bip.1974.360130103.

Masuda, A., Ushida, K., & Okamoto, T. (2005). New fluorescence correlation spectroscopy enabling direct observation of spatiotemporal dependence of diffusion constants as an evidence of anomalous transport in extracellular matrices. *Biophysical Journal, 88*(5), 3584−3591. http://dx.doi.org/10.1529/biophysj.104.048009.

Meseth, U., Wohland, T., Rigler, R., & Vogel, H. (1999). Resolution of fluorescence correlation measurements. *Biophysical Journal, 76*(3), 1619−1631. http://dx.doi.org/10.1016/S0006-3495(99)77321-2.

Ruprecht, V., Wieser, S., Marguet, D., & Schutz, G. J. (2011). Spot variation fluorescence correlation spectroscopy allows for superresolution chronoscopy of confinement times in membranes. *Biophysical Journal, 100*(11), 2839–2845. http://dx.doi.org/10.1016/j.bpj.2011.04.035.

Saxton, M. J. (2005). Fluorescence corralation spectroscopy. *Biophysical Journal, 89*(6), 3678–3679. http://dx.doi.org/10.1529/biophysj.105.074161.

Schwille, P., Haupts, U., Maiti, S., & Webb, W. W. (1999). Molecular dynamics in living cells observed by fluorescence correlation spectroscopy with one- and two-photon excitation. *Biophysical Journal, 77*(4), 2251–2265. http://dx.doi.org/10.1016/S0006-3495(99)77065-7.

Singer, S. J., & Nicolson, G. L. (1972). The fluid mosaic model of the structure of cell membranes. *Science, 175*(4023), 720–731.

Tetin, S. Y. (2013a). Methods in enzymology. Fluorescence fluctuation spectroscopy (FFS), part A. Preface. *Methods in Enzymology, 518*, xi–xii. http://dx.doi.org/10.1016/B978-0-12-388422-0.09990-8.

Tetin, S. Y. (2013b). Methods in enzymology. Fluorescence fluctuation spectroscopy (FFS), part B. Preface. *Methods in Enzymology, 519*, xiii–xiv. http://dx.doi.org/10.1016/B978-0-12-405539-1.10000-4.

Wawrezinieck, L., Lenne, P. F., Marguet, D., & Rigneault, H. (2004). Fluroescence correlation spectroscopy to determine diffusion laws: application to live cell membranes. *Proceedings of the Society Photo-Optical Instrumentation, 5462*, 92–102. http://dx.doi.org/10.1117/12.545014.

Wawrezinieck, L., Rigneault, H., Marguet, D., & Lenne, P. F. (2005). Fluorescence correlation spectroscopy diffusion laws to probe the submicron cell membrane organization. *Biophysical Journal, 89*(6), 4029–4042. http://dx.doi.org/10.1529/biophysj.105.067959.

Wenger, J., Conchonaud, F., Dintinger, J., Wawrezinieck, L., Ebbesen, T. W., Rigneault, H., … Lenne, P. F. (2007). Diffusion analysis within single nanometric apertures reveals the ultrafine cell membrane organization. *Biophysical Journal, 92*(3), 913–919. http://dx.doi.org/10.1529/biophysj.106.096586.

Yechiel, E., & Edidin, M. (1987). Micrometer-scale domains in fibroblast plasma membranes. *Journal of Cell Biology, 105*(2), 755–760.

Zacharias, D. A. (2002). Sticky caveats in an otherwise glowing report: oligomerizing fluorescent proteins and their use in cell biology. *Science's STKE, 2002*(131), pe23. http://dx.doi.org/10.1126/stke.2002.131.pe23.

CHAPTER

Investigating symmetry breaking in yeast: from seeing to understanding

2

P. Singh[1], A. Das[1], R. Li

Johns Hopkins University, Baltimore, MD, United States

[1]*Corresponding authors: E-mail: psingh23@jhmi.edu; adas16@jhmi.edu*

CHAPTER OUTLINE

Methods in Cell Biology, Volume 139, ISSN 0091-679X, http://dx.doi.org/10.1016/bs.mcb.2016.12.001

Abstract

Diverse cellular functions require the establishment of cell polarity. Although the morphological manifestation of cell polarity can be complex, common principles and conserved regulatory pathways can be studied leveraging the powerful genetic tools in the budding yeast *Saccharomyces cerevisiae*. A vast array of genetic and cell biological tools and interdisciplinary approaches have been employed to understand the establishment and maintenance of cell polarity in budding yeast. These tools enabled the identification of key molecular components, their mechanisms of action, and the pathway governing the spatiotemporal properties of cell polarity. In this review, we describe a set of useful methods with special focus on the combination of genetic manipulation and quantitative live-cell imaging, as well as advanced microscopy-based quantitative analysis, for studying symmetry-breaking process in budding yeast. Specific examples are included to illustrate the utilities of each method.

INTRODUCTION

Cell polarization is a fundamental process necessary for a variety of functions in diverse cell types such as neuronal signal transmission (Witte & Bradke, 2008), ion transport across epithelium (Drubin & Nelson, 1996), cell migration (Chung, Funamoto, & Firtel, 2001), and pollen tube growth in plants (Kost, 2008). Although cells from different lineages and organisms, even within a single animal, display vast physiological and morphological diversity, cell polarization processes share common underlying principles conserved in eukaryotes from yeast to human (reviewed in Etienne-Manneville, 2013; Li & Gundersen, 2008).

The budding yeast *Saccharomyces cerevisiae* is an attractive model for cell polarization as it exhibits distinct polarized growth that develop into a daughter cell (see Fig. 1A, D, and E) (Bi & Park, 2012; Martin & Arkowitz, 2014; Pruyne & Bretscher, 2000a, 2000b) The site of polarized bud growth at the cortex is selected in a nonrandom fashion in *S. cerevisiae*. In normal haploid yeast, buds grow in axial pattern in which both mother and daughter cells develop a bud adjacent to their previous division site. Diploid cells bud in a bipolar fashion by which mother cells form buds either axially or on opposite pole to the previous division site at cell cortex while daughter cell exclusively buds at the opposite pole (Chant & Herskowitz, 1991; Chant & Pringle, 1991; Freifelder, 1960; Hicks, Strathern, & Herskowitz, 1977). Cdc42, a small Rho GTPase and conserved regulator of cell polarity, marks the incipient bud site in *S. cerevisiae*. Cdc42 was first discovered in the budding yeast (Adams, Johnson, Longnecker, Sloat, & Pringle, 1990), subsequently its role in the cell polarization in many higher eukaryotic organisms was recognized (reviewed in Etienne-Manneville, 2013; Li & Gundersen, 2008). Budding pattern in haploid and diploid yeast are regulated by a set of genes collectively called "BUD genes" (Bi & Park, 2012; Chant & Herskowitz, 1991; Chant, Mischke, Mitchell, Herskowitz, & Pringle, 1995; Chant & Pringle, 1995; Park & Bi, 2007). These gene products determine the site of budding where Cdc42 gets recruited.

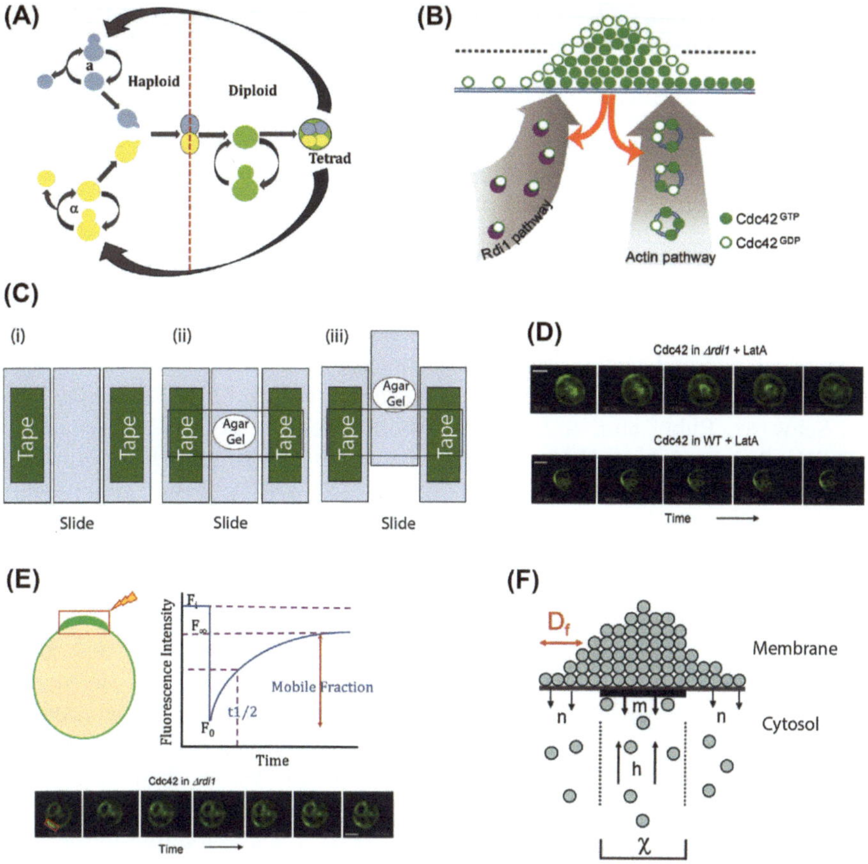

FIGURE 1 Life cycle of budding yeast and measurements of Cdc42 dynamics at the polar cortex.

(A) Life cycle of budding yeast. (B) A schematic for Cdc42 trafficking at polar cortex through actin and GDI (Rdi1)-mediated pathways. Cdc42 polarization is controlled by actin-mediated vesicular (*blue circles*) trafficking and Rdi1-dependent (*purple crescents*) pathways. (C) Preparation of agar gel pad, see text for detailed procedure. (D) Example montage of time-lapse movie showing Cdc42 polar cap in *Δrdi1* and wild-type (WT) cells upon treatment with 100 μM latrunculin A. (E) Top panels: schematic of fluorescence recovery after photobleaching (FRAP) experiment at polar cap and recovery curve (see text for details). Bottom: an example montage of FRAP for a cell expressing GFP-Cdc42 in the *Δrdi1* background. (F) Overview of the mathematical model, see text for symbol notations.

(D–F) Adapted from Slaughter, B. D., Das, A., Schwartz, J. W., Rubinstein, B., & Li, R. (2009). Dual modes of cdc42 recycling fine-tune polarized morphogenesis. Developmental Cell, 17, *823–835 with permission.*

Symmetry breaking refers to the transition from a non-polarized cell to one with a distinct asymmetry that begins to specify the axis of polarized growth or morphogenesis (Howell & Lew, 2012; Johnson, Jin, & Lew, 2011; Li & Bowerman, 2010; Slaughter, Smith, & Li, 2009). Although symmetry breaking in budding yeast is normally guided by spatial cues, such as bud scars or the pheromone gradient, disabling of these cues does not prevent symmetry breaking but results in polarization in random directions (Chant & Herskowitz, 1991; Chant et al., 1995; Chant & Pringle, 1995; Howell et al., 2009; Irazoqui, Gladfelter, & Lew, 2003; Kozubowski et al., 2008; Layton et al., 2011; Ozbudak, Becskei, & van Oudenaarden, 2005; Smith et al., 2013). A key indication of symmetry breaking in budding yeast is the localization of Cdc42 to a distinct cortical site forming a polar cap which precedes both the incipient bud and mating projection formation (Arkowitz, 1999, 2009, 2013; Butty et al., 2002; Layton et al., 2011; Savage, Layton, & Lew, 2012; Slaughter, Das, Schwartz, Rubinstein, & Li, 2009; Wedlich-Soldner, Altschuler, Wu, & Li, 2003; Wu, Savage, & Lew, 2013). Cdc42 activation as the main driver for symmetry breaking is shown by studies that artificially induced Cdc42 activation either locally or globally (O'Neill, Kalyanaraman, & Gautam, 2016; Toettcher, Weiner, & Lim, 2013; Wedlich-Soldner et al., 2003). As such, observing and manipulating Cdc42 activation and distribution have been frequent approaches in the study of yeast cell polarity and will be a focus of this review.

Like all Rho GTPases, Cdc42 cycles between inactive GDP- and active GTP-bound state (Park & Bi, 2007; Parri & Chiarugi, 2010; Sit & Manser, 2011; Van Aelst & D'Souza-Schorey, 1997). Cdc42 GTPase cycle not only modulates downstream morphogenetic events but also impacts Cdc42's localization and dynamic association with the membrane cortex (Marco, Wedlich-Soldner, Li, Altschuler, & Wu, 2007; Slaughter, Das, et al., 2009; Wedlich-Soldner et al., 2003; Wedlich-Soldner, Wai, Schmidt, & Li, 2004). Cdc42 targeting to the polar cortex is mediated by two distinct but overlapping mechanisms: (1) actin-mediated vesicular transport (Wedlich-Soldner et al., 2003, 2004) (Fig. 1B) and (2) Guanine nucleotide dissociation inhibitor (GDI)—mediated chaperoning and extraction of Cdc42 to and from the plasma membrane (PM), respectively (Das et al., 2012; Koch et al., 1997; Slaughter, Das, et al., 2009). Actin cables and type V myosin enable the transport of exocytic vesicles carrying Cdc42 to the polar cap (Evangelista et al., 1997; Marco et al., 2007; Orlando et al., 2011; Schott, Ho, Pruyne, & Bretscher, 1999; Slaughter et al., 2013; Wedlich-Soldner et al., 2003, 2004). Rdi1 is the sole RhoGDI in yeast that forms soluble complex with Cdc42 (Masuda et al., 1994). Rdi1 has been shown to extract Cdc42 from internal membranes (Eitzen, Thorngren, & Wickner, 2001) and the PM (Richman et al., 2004; Tcheperegine, Gao, & Bi, 2005). Together, actin- and GDI-mediated Cdc42 trafficking mechanisms regulate the establishment and maintenance of polarity machinery at the cortex during symmetry breaking in yeast.

To understand the establishment and maintenance of cell polarity in budding yeast, a vast array of genetic and cell biological tools and interdisciplinary approaches have been employed. These methods have not only enabled the identification of key molecular components but also their mechanism of action in regulating symmetry breaking. In this review, we describe the commonly used methods with special focus on the combination of genetic manipulation and quantitative live-cell imaging, as well as advanced microscopy-based quantitative methods, to study symmetry-breaking process in budding yeast. The utility of these methods is illustrated with specific examples. Due to space limitation, we are not able to include many other useful methods for studying yeast cell polarity.

1. GENETIC TOOLS

S. cerevisiae has heterothallic life cycle with stable haploid and diploid states that can be easily converted from one to the other; as such, phenotypes of recessive mutations can be studied conveniently (Botstein & Fink, 2011; Hinnebusch & Johnston, 2011) (Fig. 1A). Another useful characteristic of budding yeast is the high efficiency of homologous recombination that allows easy genetic manipulations in *S. cerevisiae* genome thus making *S. cerevisiae* the workhorse for research in cell biology. To enable genome modifications in a high-throughput manner, a toolkit of plasmids with choice of selectable markers has been generated for yeast (Sikorski & Hieter, 1989). Plasmids carrying gene of interest can be either introduced as self-replicating molecules or integrated into the yeast genome. Currently, genetic manipulations such as deletions, truncations, and gene tagging are commonly achieved through single-step transformation of a PCR product with both a selectable marker and partial homology (40–60 bp) to the desired locus (Gardner & Jaspersen, 2014; Longtine et al., 1998; Sheff & Thorn, 2004). Basic methods for gene manipulation in yeast, detailing the design strategies and generation of PCR products for gene deletions, point mutations, and tagging with epitopes or fluorescent proteins are discussed extensively in the literature (Botstein & Fink, 2011; Gardner & Jaspersen, 2014; Hinnebusch & Johnston, 2011; Longtine et al., 1998; Sheff & Thorn, 2004). By employing these methods, genome-wide yeast strain libraries carrying open reading frame deletions (Giaever et al., 2002; Winzeler et al., 1999), genes tagged with high-affinity epitope for biochemical purification (Ghaemmaghami et al., 2003; Puig et al., 2001) or green fluorescent protein (GFP) (Huh et al., 2003) are generated. Synthetic genetic array strategies (Baryshnikova, Costanzo, Dixon, et al., 2010; Baryshnikova, Costanzo, Kim, et al., 2010; Chong et al., 2015; Costanzo et al., 2010; Wagih et al., 2013) have been developed to reveal genetic interactions among different genes. Genomic collections enabled rapid progress in our understanding about yeast cell biology (Giaever & Nislow, 2014). Importantly, the thoroughly annotated genome of *S. cerevisiae* (Goffeau et al., 1996) has allowed the identification of gene/protein homologs in human and other eukaryotes (Zhang & Bilsland, 2011). Functional analysis of gene homologs of other eukaryotes can be performed in detail with high accuracy in budding yeast within a short span of time.

In order to study how cells break symmetry in the absence of spatial cues, genes required for bud site selection are removed. *AXL2 and RAX1* genes determine axial or bipolar budding pattern, respectively (Chant & Herskowitz, 1991; Chant et al., 1995; Chant & Pringle, 1991, 1995; Fujita et al., 2004; Roemer, Madden, Chang, & Snyder, 1996; Sanders, Gentzsch, Tanner, & Herskowitz, 1999), whereas *RSR1* is required to transmit signal to Cdc42 from upstream bud scar markers (Bender, 1993; Park & Bi, 2007; Park, Chant, & Herskowitz, 1993; Park, Kang, & Rachfal, 2002). Thus, the *Δrsr1* yeast strain was commonly used for studying spatial cue-independent symmetry breaking (Howell et al., 2009; Kozubowski et al., 2008). More recently, Rsr1, also a small GTPase, was recognized to be directly involved in symmetry breaking in addition to having a role in bud site selection (Kang, Beven, Hariharan, & Park, 2010; Kozminski et al., 2003; Park & Bi, 2007; Smith et al., 2013). Therefore, a more specific model was created by double deletion of *AXL2* and *RAX1*, leaving the Rsr1 GTPase module intact (Smith et al., 2013). In another example, deletion and point mutants were generated for Cdc42, Rdi1 and components of actin pathway to gain deeper insights into Cdc42 polarization. Conditional mutants such as *arp2-3* (Winter, Podtelejnikov, Mann, & Li, 1997), *myo2-66* (Johnston, Prendergast, & Singer, 1991) *tpm1-2* (Pruyne, Schott, & Bretscher, 1998) as well as deletion of genes whose essentiality is covered by a homolog, such as *bni1Δ* (Evangelista et al., 1997; Ozaki-Kuroda et al., 2001; Sagot, Klee, & Pellman, 2002), have been useful in determining the role of actin in vesicular transport of Cdc42 to cortex during cell polarization (Freisinger et al., 2013; Slaughter, Das, et al., 2009; Wedlich-Soldner et al., 2003; Yu, Crevenna, Bettenbuhl, Freisinger, & Wedlich-Soldner, 2011). On the other hand, *Δrdi1* and point mutants of Cdc42 such as $Cdc42^{R66E}$, non-Rdi1 interacting mutant (Slaughter, Das, et al., 2009), and $Cdc42^{Q61L}$ (GTP locked, constitutively active) and $Cdc42^{D57Y}$ (GDP locked) (Wedlich-Soldner et al., 2003, 2004; Ziman, O'Brien, Ouellette, Church, & Johnson, 1991) helped in defining the function of Rdi1 pathway in cell polarization (Das et al., 2012; Freisinger et al., 2013; Slaughter, Das, et al., 2009; Slaughter et al., 2013).

2. MICROSCOPY-BASED METHODS TO VISUALIZE CELL POLARIZATION

With the advent of genetically encoded fluorescent proteins, fluorescence microscopy has enabled researchers to visualize dynamic subcellular structures and biomolecule localizations in live cells. *S. cerevisiae* offers considerable advantage for live-cell observation with fluorescence microscopy, in which proteins in yeast can be tagged at the native loci with a fluorescent protein of choice for observation of the tagged proteins at their endogenous concentrations. The availability of arrayed genomic GFP-tagged library (Huh et al., 2003) allows researchers to readily study the intracellular localization of most yeast proteins in live cells. Recently, using additional fluorescent protein and automated single-cell analysis, localization library and database of yeast proteins in different subcellular compartments have been produced (Chong et al., 2015; Koh et al., 2015). However, because some of these libraries were constructed in high-throughput manner, it is important to validate correct tagging with genomic PCR. The functionality of the tagged genes can be indicated by the growth ability and genetic interactions with mutants with known synthetic effects. Of note, because the function of Cdc42 and other Rho family GTPases depends on C-terminal prenylation, these proteins can only be tagged at the N-termini. The N-terminal tagged Cdc42 supports the essential function at 23°C but not 37°C (Freisinger et al., 2013).

2.1 SINGLE CELL TIME-LAPSE IMAGING

Recording a cell to undergo symmetry breaking through time-lapse imaging offers detailed information on the initiation, establishment, and maintenance of polarization in time and space. For example, monitoring Cdc42 accumulation at the polar cap, which initiates symmetry breaking, provides information on the stability and strength of polarity as observed by duration and concentration of Cdc42 in the polar cap (Slaughter, Smith, et al., 2009; Slaughter et al., 2013; Smith et al., 2013). Because a large number of different proteins are known to concentrate at polar cortex during cell polarization (Gao et al., 2011), time-lapse imaging with two or more colors helps distinguish the potential function of various proteins based on their sequence of appearance at the cortex. GFP and mCherry are frequently used pairs due to their excellent spectral separation (Shaner et al., 2004); however, newer generations of red fluorescent proteins such as mKate2 (Lee, Lim, & Thorn, 2013; Shcherbo et al., 2009) may be used since mCherry has a propensity of disrupting protein function for reasons not understood.

2.1.1 Slide preparation for short-time lapse imaging

The most crucial part of live cell imaging of yeast is to keep cells stationary while they grow. For short time-lapse imaging, yeast cells can be directly mounted onto glass slide. Detailed information on the media and culture conditions for live-imaging budding yeast has been described previously (Slaughter, Unruh, & Li, 2011).

1. Freshly streak the strain of interest from a frozen glycerol stock onto a yeast extract peptone dextrose (YEPD)—agar plate. Upon incubation for 48 h at room temperature (RT) or 30°C, yeast colonies become visible. The plates can be stored for future use for ~2 weeks at 4°C.
2. Inoculate single colonies from the plate in synthetic complete (SC) or synthetic dropout (SD) media, defined growth media with minimal autofluorescence, and culture overnight at RT or 30°C, shaking at 200 rpm.
3. After overnight growth, reinoculate the yeast culture into fresh SC or SD media at an OD of 0.1—0.2 and grow for another 3—4 h. Subsequently, transfer ~4 μL cell culture directly onto the glass slide for imaging. Alternatively, concentrate 1 mL of the yeast culture by centrifugation at low speed (~4000 rpm) and

resuspend the cell pellet in 50–100 μL before pipetting ~4 μL onto a microscope slide (see Note 1).

4. Gently cover the cell suspension on the microscope slide with a 22 × 22 mm, no. 1.5 (0.16- to 0.19-mm), cover glass. To immobilize cells on the slide, turn the slide over and press downward gently onto clean tissue paper to squeeze out the excess media. Image the prepared slides for up to 20 min (see Note 2). For longer imaging experiments, see in the following paragraph.

2.1.2 Slide preparation for long time-lapse imaging

1. Prepare 1% agarose in 1 mL of SC or SD media. Subsequently, heat agarose mix in SC or SD media at 75°C for 5–10 min and gently invert the tubes a couple of times. The hot agarose mix can be kept at 65°C for 3–4 h (see Notes 3,4).
2. In the meantime, add some thickness to two different slides by putting a piece of regular adhesive tape on the flat surface of the slide. Place a clean slide between these two taped-slides side by side in a row (as shown in Fig. 1C, (i)).
3. Cool the homogenous solution of agarose for few minutes and drop ~50 μL of hot agar in the middle of the center slide avoiding air bubble formation.
4. Place another clean slide on top of the agar drop across the center slide, such that it rests on the taped-slides (see Fig. 1C, (ii)). Press down top slide gently to make a gel sandwich between the two slides of the thickness of the tape. Let the gel sandwich cool at RT for 3–5 min. To prevent dehydration, leave the top slide in place until cells are ready for mounting (see Note 4).
5. Meanwhile, put a 1 mL tube of valap (see notes) in the 100°C heat block to melt (see Note 5).
6. Carefully separate the two slides of the gel sandwich by gently pushing the center slide parallel to other two taped-slides, such that the top slide remains resting on the two taped-slides (see Fig. 1C, (iii)). The agar gel pad generally stays on the bottom-center slide (see Note 6).
7. Transfer 2–3 μL of concentrated cell culture, prepared as aforementioned, onto the agarose pad. Put a clean 22 × 22 mm, no. 1.5 (0.16- to 0.19-mm) cover glass on top and apply even pressure gently to immobilize the cells between the pad and cover glass (see Notes 6, 7).
8. Finally, seal the cover glass to the slide by first adding ~5–10 μL of hot liquid valap to each corner of the coverslip to fix it in place, then quickly applying a fine layer of valap along the edges of the coverslip.

Time-lapse microscopy of live yeast cells, using the methods described earlier, has led to many discoveries regarding the molecular dynamics and distributions during cell polarization. For example, time-lapse imaging of fluorescently tagged Cdc42 showed that the polar cap is less stable in the *Δrsr1* background than the *Δaxl2 Δrax1* strain particularly in the absence of actin-mediated Cdc42 transport. These results indicate that Rsr1 not only carry information for bud site selection but also participate in cell polarization process in general (Smith et al., 2013). Moreover, through the careful analysis of single-cell, time-lapse movies, it was discovered that actin-based transport and GDI-mediated recycling of Cdc42 are important for the polarization of Cdc42 during bud formation (Freisinger et al., 2013; Slaughter, Das, et al., 2009; Slaughter et al., 2013; Smith et al., 2013). Inhibiting actin polymerization in wild-type (WT) cells significantly reduces the Cdc42 polarization (Wedlich-Soldner et al., 2004). Furthermore, Cdc42 fails to polarize in *Δrdi1* cells upon inhibiting actin polymerization with the drug latrunculin A (LatA) (Freisinger et al., 2013; Slaughter, Das, et al., 2009; Slaughter et al., 2013; Smith et al., 2013). These observations show that Rdi1 and actin represent two pathways for Cdc42 localization to the cell cortex. Cells polarize as long as one of these pathways is functional but fail to polarize when both are inhibited (Fig. 1D) (Freisinger et al., 2013; Slaughter, Das, et al., 2009; Slaughter et al., 2013; Smith et al., 2013).

2.2 BULK CELL POLARIZATION ASSAYS

An easier method of quantifying the efficiency of cell polarization in budding yeast than making a large number of movies is to arrest cells in G1 cell cycle phase, the release from which allows observation of synchronized polarization in a large number of cells. Polarization occurs soon after release from the arrest. Several methods for G1 cell cycle arrest in budding yeast are described in the literature (Breeden, 1997; Manukyan, Abraham, Dungrawala, & Schneider, 2011). Among them, pheromone-mediated arrest is the most efficient and commonly used method to arrest cells in G1 phase (Breeden, 1997; Manukyan et al., 2011). Haploid yeast exists as in *MATa* and *MATα* mating forms. *MATα* cells secrete a 13-amino acid pheromone called α-factor that arrests *MATa* cells in the G1 phase. α-factor can be obtained commercially and directly applied to arrest *MATa* cells. Successful arrest can be visually confirmed by the presence of large, unbudded cells with "shmoo"-like shape. Shorter arrest time and lower concentration of the α-factor can minimize the cell shape distortion while still achieving cell cycle arrest, but these parameters should be determined empirically for specific strain or strain background used. Synchronous cell polarization can be achieved by releasing α-factor–arrested cells into fresh medium as explained in the following steps.

1. Grow a yeast culture overnight, then refresh it the next day by diluting it to an OD of ~0.05 and allow it to grow for 3–4 h (~2 cell cycles).
2. Once the cell culture reaches exponential growth, at an OD of 0.2–0.3, add α-factor at ~0.15 μg/mL (see Notes 8, 9).
3. After 90–120 min, observe the morphology of the cells under a microscope to check for growth arrest. G1 cells should be unbudded and some may form shmoos (see Note 10).
4. To release from the G1 arrest, thoroughly wash the cells three times in quick succession with cold sterile water by spinning them down at high speed followed by vacuum aspiration (see Note 11). These washes also disrupt polarization caused by the pheromone and thus reset the polarity state.

5. After removing traces of α-factor, release the cells into YEPD media of 1/4 to 1/2 volume of the original culture at an OD of 0.4–0.8. Budding should be observed to initiate at ~10 min, samples can be collected at regular intervals such as 20, 50, and 80 min after release.
6. Fix GFP expressing cells in 4% paraformaldehyde for 15 min before washing in phosphate buffered saline whereas cells should be fixed for longer time >30 min for staining with phalloidin or antibodies. Fixed cells are imaged by directly mounting onto glass slide as described earlier; they can be stored at 4°C for up to 48 h before imaging (see Note 12).

Cell polarization efficiency at each time point is quantified by counting cells with a polarized cortical accumulation of Cdc42 or other polarity markers at different time points after release from G1 arrest. To monitor the role of actin-mediated transport in polarization, cell polarization assay can be performed by releasing G1 arrested cells into media containing LatA at a concentration of ~50–100 μM (see Note 13). A significant reduction in polarizing cells was observed under LatA treatment suggesting that actin is required for efficient Cdc42 polarization (Butty et al., 2002; Gulli et al., 2000; Smith et al., 2013). Moreover, using this assay to observe *Δrsr1* and *Δaxl2 Δrax1* under LatA treatment showed that the polarization rate of *Δaxl2 Δrax1* cells was similar to WT cells, whereas *Δrsr1* cells treated with LatA polarized with significantly reduced efficiency (Smith et al., 2013), confirming that the Rsr1 GTPase has additional functions in cell polarization beyond bud site selection.

2.3 FLUORESCENCE RECOVERY AFTER PHOTOBLEACHING

Although single-cell imaging and polarization assays are useful in capturing the sequential order and kinetics of polarization of specific factors to the cell cortex, these methods do not provide information on the molecular dynamics of individual proteins, such as diffusion and rates of association with and dissociation from the membrane. Fluorescence recovery after photobleaching (FRAP) is a relatively straightforward microscopy-based approach that can provide information on dynamics of individual molecules in living cell using GFP-tagged strains. In this technique, fluorescently tagged molecules within a region of interest (ROI) are first irreversibly bleached with an intense laser beam. Thereafter, fluorescence recovery in the region is recorded at low laser power. Fluorescence recovery occurs as unbleached fluorescent molecules enter and bleached, dark molecules exit the ROI. Thus, the rate of fluorescence recovery is an indication of the diffusion rate, as well as the off rate for proteins bound to the membrane or cortex. Here, we provide some practical guidelines for optimal FRAP analysis in yeast.

2.3.1 *Practical considerations for fluorescence recovery after photobleaching in* Saccharomyces cerevisiae

1. *Fluorescent probes*: One of the key prerequisite of successful FRAP experiments is the appropriate fluorophore. Although many genetically encoded fluorescent

proteins have been developed (Shaner et al., 2004), GFP has been the fluorescent protein of choice due to its photostability and high quantum yield (Day & Davidson, 2009; Loren et al., 2015; Prendergast, 1999; Yang, Moss, & Phillips, 1996). Proteins tagged with GFP can be visualized in cells with low laser power over time with little photobleaching during acquisition thus allowing accurate quantification of fluorescence recovery (Hirschberg et al., 1998; Lippincott-Schwartz, Roberts, & Hirschberg, 2000).

2. *Photobleaching Conditions*: The photobleaching time for FRAP experiments is generally considered to be instantaneous. However, in practice, the true time required to achieve adequate bleaching varies depending on the laser power and extent of fluorophore brightness. For proteins that display rapid recovery, an overly long bleach time would obscure the initial recovery points, leading to an underestimation of the diffusion coefficient and mobile fraction; thus, high laser power should be used for a short duration to bleach these samples. If, however, the sample is sensitive to high laser power, bleaching should be performed at low laser power for longer bleach time. As a rule of thumb, the bleach time should be less than 5% of the characteristic recovery time for each protein (Loren et al., 2015; Meyvis, De Smedt, Van Oostveldt, & Demeester, 1999). Additionally, the laser power used for image acquisition should be minimized during the recovery phase of FRAP experiments to prevent further photobleaching, and a nonbleached sample should be imaged under the same conditions as a control.
3. *FRAP analysis*: After subtracting background intensity, fluorescence intensity at different time points of FRAP experiment (both pre- and post-bleach) is normalized to prebleach intensity to generate a FRAP curve as shown in Fig. 1E. The mobile fraction (Mf) is the percent of diffusible molecules, determined based on the extent recovery of the bleached fluorescent signal, while the $t_{1/2}$ corresponds to the amount of time needed to reach half of the maximum fluorescence recovery. Based on experimental condition and fitting parameters a FRAP recovery trace is subjected to appropriate model to determine molecular diffusion coefficient (D). F_i and F_0 signify fluorescence intensity at pre- and post-bleach, whereas F_∞ refers to fluorescence intensity at steady-state after recovery.

$$F(t) = F_\infty(1 - e^{\tau \cdot t}) \tag{1}$$

$$t_{1/2} = \ln 0.5/\tau \tag{2}$$

$$Mf = (F_\infty - F_0)/(F_i - F_0) \tag{3}$$

For freely diffusing molecules, FRAP recovery curve is fit to simple exponential function as in shown Eq. (1), Mf and $t_{1/2}$ are calculated using Eqs. (2) and (3). Inverse FRAP (iFRAP), a derivative of FRAP, is performed much like normal FRAP. In iFRAP, whole cell except the ROI is photobleached and instead of recovery, loss of fluorescence is observed over time. Using iFRAP, dissociation rate of

Cdc42 from PM was determined (Das et al., 2012; Slaughter, Das, et al., 2009; Slaughter et al., 2013). Although Cdc42 appears to form a stable accumulation at the cell cortex in polarized cells (see Fig. 1), FRAP analysis revealed that individual molecules of Cdc42 are highly dynamic at the polar cap (Das et al., 2012; Freisinger et al., 2013; Marco et al., 2007; Slaughter, Das, et al., 2009; Slaughter et al., 2013; Wedlich-Soldner et al., 2004). Cdc42 dynamics at yeast cell cortex is mediated through actin-based vesicular trafficking and endocytosis and Rdi1-dependent recycling. The *Δrdi1* strain showed slow recovery of Cdc42 at the polar cap after bleaching, indicating that Rdi1 is responsible for fast Cdc42 recycling that contributes to the dynamic maintenance of Cdc42 at the cortex. In the absence of Rdi1 recycling, Cdc42 dissipates laterally to the cell cortex from polar cap diminishing its polarization. As such, Rdi1 plays a positive role (Slaughter, Das, et al., 2009) in cell polarization, rather than having purely an inhibitory function (Cole, McLaughlin, & Johnson, 2007; Hoffman, Nassar, & Cerione, 2000; Koch et al., 1997; Masuda et al., 1994; Richman et al., 2004; Tiedje, Sakwa, Just, & Hofken, 2008). In the absence of Rdi1-mediated fast recycling of Cdc42, there is an increase in the number of yeast cells with dual buds (Freisinger et al., 2013), indicating that the GDI pathway is required to focus Cdc42 into a single polarization site.

2.3.2 Mathematical modeling of the fluorescence recovery after photobleaching data

Although FRAP measures Cdc42 recycling rate, it does not measure parameters describing individual dynamic events such as diffusion of Cdc42 in the PM, its rate of internalization from or transport to the polar cortex. For this, mathematical models were developed for the analysis of FRAP data, which helped understand the contribution of different dynamic processes to the morphogenetic outcome of yeast cell polarization (Marco et al., 2007; Slaughter, Das, et al., 2009). The mathematical model Eq. (4) was developed to explain steady-state Cdc42 distribution on the polarized yeast PM as a result of dual recycling mechanisms (Slaughter, Das, et al., 2009), based on the earlier work describing single recycling pathway (Marco et al., 2007).

$$\frac{\partial f}{\partial t} = D_f \Delta f - m_\chi f - n(1-\chi)f + h_\chi F_c \tag{4}$$

The framework of the mathematical model includes; concentric overlapping Cdc42 delivery window (χ) of actin and Rdi1 pathway with delivery rate h, internalization rate m inside the window and n outside, and Cdc42 membrane diffusion (D_f) (Fig. 1F). Each of these parameters is sum of the actin and Rdi1-mediated Cdc42 recycling pathway. A dynamic equilibration between Cdc42 diffusion D_f, internalization m, n and retargeting back to the polar cortex h is critical for maintaining the steady-state Cdc42 concentration on the polar cortex. Where $f(r,\phi,t)$ denotes Cdc42 concentration on yeast surface at coordinate (r,ϕ) at any given time point (t), $D_f\Delta f$ accounts for Cdc42 diffusion along the membrane. F_c denotes amount of cytosolic (internal) Cdc42; χ is a window function and equals 1 inside the delivery window and equals 0 outside the window. At steady-state the derivative vanishes $\left(\frac{\partial f}{\partial t} = 0\right)$, Eq. (4) converts into the form (5)

$$D_f \Delta f - m_\chi f - n(1-\chi)f + h_\chi F_c = 0 \tag{5}$$

Using FRAP data, image analysis, and published D_f value of 0.036 $\mu m^2/s$ (Marco et al., 2007) model parameters *m*, *n*, and *h* could be extracted by numerical fitting using equations derived from Eq. (4) for single cells. Then, these parameters are combined with the analytical solution of Eq. (5) to predict the Cdc42 cortical distribution, which can be compared with that measured for the corresponding cell prior to the FRAP experiment. Excellent agreement between the two validated the model. The potential impact of Cdc42 distribution is the shape of the polarized growth regulated by Cdc42, which dictates the morphogenetic outcome. Although the continuous steady-state model nicely predicts a stable, polarized distribution of Cdc42, it was insufficient to describe the system when membrane trafficking was accounted for explicitly (Layton et al., 2011; Savage et al., 2012). A later study using high-resolution imaging and discrete numerical simulation revealed existence of lipid microdomains in the PM, in which Cdc42 exhibits a different diffusion coefficient relative to elsewhere in the cortex. Model simulation with this feature demonstrated its importance for maintenance of the stable polar cortex (Jose, Tollis, Nair, Sibarita, & McCusker, 2013; Slaughter et al., 2013). Such lipid microdomains also found to be essential for stable polar cortex in fission yeast (Makushok, Alves, Huisman, Kijowski, & Brunner, 2016). In addition to aforementioned, mathematical modeling in combination with live cell imaging in yeast have proven to be a powerful approach to understand role of key determinants and requirements of positive feedback loops in achieving a robust singular polarization event (Brandman, Ferrell, Li, & Meyer, 2005; Howell et al., 2009; Klunder, Freisinger, Wedlich-Soldner, & Frey, 2013; Thompson, 2013; Wedlich-Soldner et al., 2003).

2.4 FLUORESCENCE CORRELATION SPECTROSCOPY

Fluorescence correlation spectroscopy (FCS): FRAP provides measurements of the rates of molecular diffusion and exchange at the polar cortex but not concentration and dynamic interactions in the cytosol, such as the interaction of Cdc42 with Rdi1. Point FCS with a single fluorescent species is uniquely suited for accurate measurements of absolute protein concentration and average diffusion time for mobile molecules at native concentrations in live cells (Altan-Bonnet & Altan-Bonnet, 2009; Kim, Heinze, & Schwille, 2007; Slaughter, Schwartz, & Li, 2007).

2.4.1 Requirements for fluorescence correlation spectroscopy

1. Experimental set-up for FCS in live yeast requires both confocal imaging and FCS modalities, through which a laser beam is used to illuminate the focal volume in the location of interest such as cytosol (Fig. 2A, (i), see Note, 14).
2. Fluorophore is excited for several second and emission intensity trace is recorded through the single-photon—counting avalanche photodiodes.

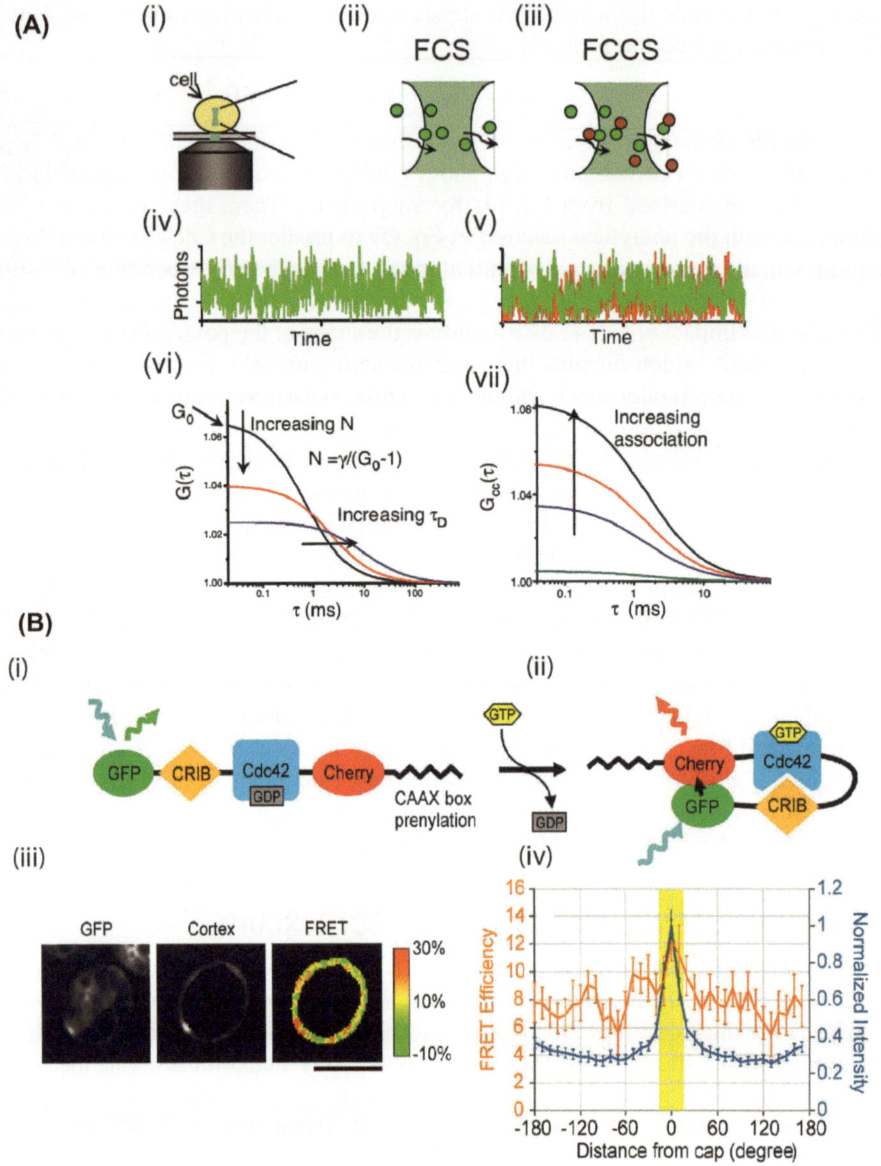

FIGURE 2 Methods to probe Cdc42 interaction and activation.

(A) (i) Schematic representation of laser excitation of the focal volume inside a cell in fluorescence correlation spectroscopy (FCS), (ii) diffusion of a green fluorophore-tagged protein (GFP) in and out of the focal volume, and (iii) codiffusion of red and green-tagged proteins through the focal volume. Fluorescence intensity traces for (iv) single channel FCS experiment, and (v) dual channel fluorescence cross-correlation spectroscopy experiment. (vi and vii) Simulated autocorrelation curves extracted from intensity traces, number of

3. Diffusion of fluorescent molecules in and out of focal volume produces the fluctuation in emission intensity (Fig. 2A, (ii–v)), which are subjected to autocorrelation function to generate an autocorrelation curve (Fig. 2A, (vi)) (Kim et al., 2007; Schwille, 2001; Slaughter & Li, 2010).
4. The autocorrelation curve thus generated can be fit to different equations to extract molar concentration from the amplitude of the curve (G_0) and the diffusion time (τ_D). The details of FCS curve fitting and considerations could be found in these references (Haustein & Schwille, 2007; Slaughter et al., 2007).
5. The average number of fluorophore (N) in the focal volume is calculated using Eq. (6) (Fig. 2A, (vi)):

$$N = \gamma / G_0 \tag{6}$$

G_0, amplitude of correlation function at zero lag time and τ_D, average diffusion time through the focal volume are obtained from fitted autocorrelation curve. γ, denotes focal volume shape factor with theoretical value of 0.36 for one photon excitation (Thompson, 1991). With N, fluorophore concentration could be measured by measuring focal volume (see Note 15).

6. Direct measurement of the focal volume is possible by 3D imaging of 100 nm fluorescent bead as described by Slaughter et al. (2013).
7. A calibrated imaging of cells expressing monomeric fluorophore and the fluorophore-tagged test protein in combination with FCS analysis with identical optical setting could provide stoichiometry of the protein (Slaughter et al., 2013).

Diffusion time and protein concentration measurement by FCS could provide important clues for its molecular interactions in the cytosol. In case of budding yeast, cytosolic Cdc42 was found to be distributed in two diffusion pools; one

mobile particles in the focal volume derived from the inverse of the autocorrelation amplitude, and diffusion time (τ_D). (vii) Determination of protein association by codiffusion from the amplitude of the cross-correlation curve. (B) Schematic representation of the fluorescence resonance energy transfer (FRET)–based biosensor for Cdc42 activation. (i and ii) GTP-GDP exchange produces $Cdc42^{GTP}$ that binds the CRIB domain, bringing the flanking GFP and mCherry into closing proximity resulting in FRET. (iii) A typical wild-type cell expressing the biosensor containing wild-type Cdc42. The image shows the FRET efficiency at each pixel within the cell as indicated by the *heat bar.* (iv) Mean FRET efficiency (*orange curve*) was measured and plotted along the cortex. Mean normalized GFP intensity at the cortex (*blue curve*) was also measured and plotted.

(A) Adapted from Slaughter, B. D., Schwartz, J. W., & Li, R. (2007). Mapping dynamic protein interactions in MAP kinase signaling using live-cell fluorescence fluctuation spectroscopy and imaging. Proceedings of the National Academy of Sciences of the United States of America, 104, *20320–20325 with permission. (B) Adapted from Smith, S. E., Rubinstein, B., Mendes Pinto, I., Slaughter, B. D., Unruh, J. R., & Li, R. (2013). Independence of symmetry breaking on Bem1-mediated autocatalytic activation of Cdc42.* The Journal of Cell Biology, 202, *1091–1106 with permission.*

predominantly fast and a minor slow diffusing pool. Later with calibrated imaging and FCS analysis, it was shown that the slow diffusion originates from Cdc42-containing exocytic vesicles (Slaughter et al., 2013). FCS study also found Rdi1 diffusion is fast with similar diffusion time to the fast component of Cdc42. This suggests that Cdc42 fast diffusion could rely on Rdi1 in the cytosol. Indeed, deletion of Rdi1 shifted Cdc42 diffusion mostly in the slow diffusing pool while overexpression of Rdi1 shifted Cdc42 predominantly in the fast diffusing pool (Das et al., 2012; Slaughter, Das, et al., 2009) suggesting existence of Cdc42-Rdi1 complex in the cytosol.

2.4.2 Fluorescence cross-correlation spectroscopy

Fluorescence cross-correlation spectroscopy (FCCS), a derivative of FCS can be employed to examine interactions of two mobile species labeled with different fluorophores. For FCCS in live yeast cells, GFP and mCherry have been successfully used as a pair due to their excellent spectral separation, though the minimal amount of overlap must be accounted for the analysis (see in the later discussion) (Bacia & Schwille, 2007; Slaughter, Das, et al., 2009; Slaughter et al., 2007).

2.4.3 Requirements and analysis of fluorescence cross-correlation spectroscopy

1. A crucial consideration for FCCS is the positive and negative control for live-cell measurement. A good strategy for such control is explained in Slaughter et al. (2007), where for positive control two fluorescent proteins, GFP and mCherry were tagged at the COOH terminus of a nonessential cytosolic protein Bat2, while for negative control a diploid yeast strain was used with Bat2-GFP in one *BAT2* locus and mCherry replacing *BAT2* in the other locus.
2. For FCCS, a pair of test proteins is tagged with two different fluorophores and intensity fluctuation is correlated between the two channels. Complex formation is indicated by protein codiffusion through the focal volume causing similar intensity fluctuation over time (Fig. 2A, (iii,v)).
3. Similar to FCS, the intensity traces for each channel are subjected to autocorrelation function to generate autocorrelation curves. A cross-correlation curve is also calculated from the fluorescence traces of the differentially labeled molecular species. The amplitude of the cross-correlation curve is indicative of the degree of complex formation between the two molecular species (Fig. 2A, (vii)). The molecular number in the focal volume for each species is calculated from the individual autocorrelation curve and the number of the bound complex can be derived from the amplitude of cross-correlation curve. With these numbers, fraction of bound species relative to the total could be derived using Eq. (7) (Rigler et al., 1998).

$$N_{bound} = N_{GT}(N_{RT} + Q \cdot N_{GT})/N_{CC} - N_{GT} \cdot Q \qquad (7)$$

N_{GT}, and N_{RT} are the number of green and red particles in the respective focal volume obtained from inverse of initial amplitude of the fitted correlation curves (see Eq. (6)). $N_{CC,}$ is the number of codiffusing molecule similarly obtained from the inverse of fitted cross-correlation curve. N_{bound}, represents fraction of red or green molecule relative to the total cytosolic population that is bound to its counterpart. Bleed through (Q) between the channels could be determined as discussed earlier (Slaughter et al., 2007).

4. Correction for bleed through could be avoided by using a recent laser switching FCCS technique, where each laser channel is excited alternatively at fastest possible frequency (Takahashi et al., 2008). However, this method is limited to the speed of the switching and hence applicable for slow diffusing components of in-vivo samples. Detailed protocol could be found, at http://research.stowers.org/imagejplugins/laser_switching_protocol.html.

Using FCCS experiment on yeast expressing GFP-Cdc42 and Rdi1-mCherry, it was observed that more than 60% cytosolic Cdc42 is complexed with Rdi1 in agreement with its fast diffusion (Slaughter, Das, et al., 2009). FCCS also revealed that the Rdi1 interaction with Cdc42 depends on its GTPase cycle, as mutant locked in GDP ($Cdc42^{D57Y}$) or GTP ($Cdc42^{Q61L}$) exhibit significantly reduced interaction (Slaughter, Das, et al., 2009). Hence, FCS-based methods are useful tools for determining protein concentration, diffusion and molecular interactions at native concentrations inside live cell.

2.5 FLUORESCENCE RESONANCE ENERGY TRANSFER

FCCS is generally limited to fast mobile molecular species and does not discriminate between direct or indirect interactions. Fluorescence Resonance Energy Transfer (FRET) is a complementary approach for detecting direct interactions between two molecular species (Jares-Erijman & Jovin, 2006; Sekar & Periasamy, 2003) based on their proximity ($< \sim 10$ nm) that is not restricted by molecular mobility. As FRET is an imaging method, it provides information on protein interactions in specific subcellular compartments or within large protein assemblies (Goley, Rodenbusch, Martin, & Welch, 2004; Mc Intyre et al., 2007). There are a number of methods for detecting FRET (reviewed in Broussard, Rappaz, Webb, & Brown, 2013; Jares-Erijman & Jovin, 2006) but we will limit our discussion to acceptor photobleaching method as it is easy to perform with relatively straightforward analysis (Slaughter et al., 2011; Smith et al., 2013; Van Munster, Kremers, Adjobo-Hermans, & Gadella, 2005). In addition, we would discuss the design of an FRET-based sensor for measuring Cdc42 activation in live yeast cells (Smith et al., 2013). FRET involves a nonradiative energy transfer between excited state donor to acceptor fluorophore situated in close vicinity (typically <10 nm). Photobleaching of the acceptor fluorophore thus results in dequenching measured by an increase in donor fluorescence. Briefly, two main requirements need to be fulfilled for successful FRET detection via acceptor bleaching. First, FRET requires a suitable fluorophore pair with overlapping emission (donor) and excitation (acceptor) spectra, such as GFP/mCherry or mTurquoise/YFP (Katta et al., 2015; Shaner et al., 2004).

Second, the microscope system must have the capability to bleach the acceptor molecule fluorescence.

2.5.1 Design of fluorescence resonance energy transfer biosensor for the guanine nucleotide exchange factor activity toward Cdc42

A FRET-based biosensor was designed to spatially interrogate the guanine nucleotide exchange factor (GEF) activity for Cdc42 activation (Smith et al., 2013). The biosensor design was similar to an earlier mammalian cell biosensor for the same purpose (Itoh et al., 2002) with some modifications. The biosensor consists of a linked construct of yeast Cdc42 with the Cdc42/Rac-interactive binding (CRIB) domain from Cla4 (Fig. 2B), which interacts only with the active, GTP-bound form of Cdc42. The Cdc42-CRIB chimera is flanked by GFP and mCherry, such that when the CRIB and $Cdc42^{GTP}$ within the sensor are bound, the GFP and mCherry are brought into sufficiently close proximity for energy transfer to occur, as illustrated in Fig. 2B. The polybasic-CAAX box region of Cdc42 was moved C-terminally to mCherry to allow proper prenylation and membrane anchorage.

2.5.2 Fluorescence resonance energy transfer experiments with the Cdc42 activation biosensor

1. Look for cells with desired polarized state in the GFP channel, and then bleach mCherry in the whole cell using intense 568 nm laser beam. Cells are then imaged in GFP channel at maximum speed with 488-nm laser with low laser power to minimize acquisition photobleaching (see Notes 16, 17).
2. As for all FRET application, positive and negative controls are highly valuable. In the case of Cdc42 activation biosensor, the control constructs were generated by introducing point mutations into the Cdc42 portion of the biosensor, resulting in either a constitutively GTP-bound (Q61L) or GDP-bound (D57Y) state.

Bem1 was proposed to mediate a positive feedback loop promoting local activation and accumulation of Cdc42 due to Bem1's ability to bind both Cdc42 and Cdc24, the Cdc42 GEF (Butty et al., 2002; Kozubowski et al., 2008). Using the FRET-based Cdc42 activation biosensor, it was shown that Bem1 boosts the GEF activity to promote high-level Cdc42 activation. This activity becomes essential for polarization in the absence of actin-based transport (Smith et al., 2013).

CONCLUSION AND FUTURE PERSPECTIVE

Significant progress has been made not only in elucidating the key molecular players and individual pathways involved in cell polarization but also in defining the fundamental principles governing cellular symmetry breaking. Studies in yeast have made important contributions to both these fronts owing to the powerful coupling of genetic tools and successful implementation of the large number of microscopy-based methods. Live-cell measurements were further coupled with mathematical modeling and computational simulations to test and formulate hypothesis on highly quantitative levels. Here, we describe basic methods and their usefulness for interrogating the cell polarity system. While the methods described will continue to bear fruit in detailed dissection of molecular and cellular mechanisms underlying cell polarization in yeast, emerging technologies, such as superresolution microscopy (Galbraith & Galbraith, 2011; Huang, Babcock, & Zhuang, 2010; Huang, Bates, & Zhuang, 2009; Leung & Chou, 2011; Schermelleh, Heintzmann, & Leonhardt, 2010), optogenetics (Jost & Weiner, 2015; Levskaya et al., 2005; Tischer & Weiner, 2014; Toettcher et al., 2013; Weitzman & Hahn, 2014; Yang, Jost, Weiner, & Tang, 2013), and microfluidics/micropatterning (Jo, Liu, Gu, Dang, & Qin, 2015; Lee, Avalos Vizcarra, Huberts, Lee, & Heinemann, 2012; Liu, Young, & Acar, 2015; Xie et al., 2012; Zhang et al., 2012) have further empowered this model system. We envision that cell polarity research in yeast incorporating new technologies will help answer some of the remaining mechanistic questions and continue to lead the field in achieving systems-level understanding of this fundamental process of biology.

NOTES

1. Media volume to resuspend cells is determined as per size of the cell pellet, ~20 μL is sufficient for small pellet whereas 100–200 μL media can be used for large pellet before mounting cells onto the slide.
2. Clean glass slide and coverslip leaving no dust particle on slide as that would add background noise and prevent a tight seal between the slide and coverslip.
3. Check the color of agar solution before use, as continuous heating at 65°C caramelizes it over time. Brown agar is unsuitable for imaging.
4. Do not use agar pads for performing FRAP, photoactivation, or two photon imaging experiments, as laser can melt the pad due to local heating, resulting in loss of focus.
5. Valap is a 1:1:1 mixture of vaseline, lanolin, and paraffin. Add ingredients in sequence to a beaker on a hot plate, allowing each to melt before adding the next. Make 1 mL aliquots and store at RT. Since the valap hardens quickly at RT, keep it at 100°C continuously before use.
6. It is best to use freshly made agar pads and pay attention to gel dehydration as it results in focal drift in Z.
7. Perfect or definite focus mechanism can be used to circumvent the sample drifting problem during long time-lapse imaging.
8. Time and concentration of α-factor used is optimized for WT cells of S288C genetic background. Mutants and strains of different genetic background may require different time and α-factor concentration to achieve cell cycle arrest.
9. To reduce the cost of α-factor, *BAR1* protease might be deleted to achieve G1 arrest at much lower concentration (~0.01 μg/mL) of α-factor.

10. To accurately count G1 arrested cells, cells may be sonicated briefly to distinguish arrested unbudded cells from cycling budded cells after α-factor treatment.

11. Cell pellets must be washed thoroughly with cold water to remove traces of α-factor. Filtration and vacuum aspiration are the fastest and most effective methods. However, great care should be taken while filtering or aspirating the media to avoid dehydrating the cell pellet.

12. Since fixation process can reduce GFP fluorescence, establish the protocol by testing different fixation times and compare fluorescence before and after fixation. Minimal fixation time is desired to stop the cell cycle without substantial reduction of GFP fluorescence.

13. LatA treatment conditions to cause loss of F-actin should be validated by actin staining with fluorescently labeled phalloidin.

14. Adjust correction collar of objective lens, and use slides of identical thickness (measured with a micrometer) to yield best results for FCS/FCCS experiments.

15. Generally, multiple autocorrelation curves are averaged for analysis, intensity traces with photobleaching possibly due to particle immobility, as judged by visual inspection, are eliminated from the analysis.

16. Laser power and exposure time should be adjusted depending on biosensor expression level in the specific cell.

17. Check mCherry fluorescence after acquisition to ensure complete photobleaching. Also, it is important to confirm that there is no bleaching of GFP upon bleaching with the 568 nm laser.

ACKNOWLEDGMENTS

The authors thank K. Swaney from Johns Hopkins School of Medicine and S. Smith, J. Unruh, B. Slaughter, and B. Rubinstein from Stowers Institute for Medical Research for critically reading the manuscript. We apologize to many researchers whose work is not cited or not cited fully due to space constraints. This work is supported by the grant R35 GM118172 from the National Institutes of Health to R. Li.

REFERENCES

Adams, A. E., Johnson, D. I., Longnecker, R. M., Sloat, B. F., & Pringle, J. R. (1990). CDC42 and CDC43, two additional genes involved in budding and the establishment of cell polarity in the yeast *Saccharomyces cerevisiae*. *The Journal of Cell Biology, 111*, 131–142.

Altan-Bonnet, N., & Altan-Bonnet, G. (2009). Fluorescence correlation spectroscopy in living cells: A practical approach. *Current Protocols in Cell Biology* (Chapter 4):Unit 4 24.

Arkowitz, R. A. (1999). Responding to attraction: Chemotaxis and chemotropism in *Dictyostelium* and yeast. *Trends in Cell Biology, 9*, 20–27.

Arkowitz, R. A. (2009). Chemical gradients and chemotropism in yeast. *Cold Spring Harbor Perspectives in Biology, 1*, a001958.

Arkowitz, R. A. (2013). Cell polarity: Wanderful exploration in yeast sex. *Current Biology, 23*, R10–R12.

Bacia, K., & Schwille, P. (2007). Practical guidelines for dual-color fluorescence cross-correlation spectroscopy. *Nature Protocols Other Titles: Protocols, 2*, 2842–2856.

Baryshnikova, A., Costanzo, M., Dixon, S., Vizeacoumar, F. J., Myers, C. L., Andrews, B., & Boone, C. (2010). Synthetic genetic array (SGA) analysis in *Saccharomyces cerevisiae* and *Schizosaccharomyces pombe*. *Methods in Enzymology, 470*, 145–179.

Baryshnikova, A., Costanzo, M., Kim, Y., Ding, H., Koh, J., Toufighi, K., Youn, J. Y., … Myers, C. L. (2010). Quantitative analysis of fitness and genetic interactions in yeast on a genome scale. *Nature Methods, 7*, 1017–1024.

Bender, A. (1993). Genetic evidence for the roles of the bud-site-selection genes BUD5 and BUD2 in control of the Rsr1p (Bud1p) GTPase in yeast. *Proceedings of the National Academy of Sciences of the United States of America, 90*, 9926–9929.

Bi, E., & Park, H. O. (2012). Cell polarization and cytokinesis in budding yeast. *Genetics, 191*, 347–387.

Botstein, D., & Fink, G. R. (2011). Yeast: An experimental organism for 21st Century biology. *Genetics, 189*, 695–704.

Brandman, O., Ferrell, J. E., Jr., Li, R., & Meyer, T. (2005). Interlinked fast and slow positive feedback loops drive reliable cell decisions. *Science, 310*, 496–498.

Breeden, L. L. (1997). Alpha-factor synchronization of budding yeast. *Methods in Enzymology, 283*, 332–341.

Broussard, J. A., Rappaz, B., Webb, D. J., & Brown, C. M. (2013). Fluorescence resonance energy transfer microscopy as demonstrated by measuring the activation of the serine/threonine kinase Akt. *Nature Protocols Other Titles: Protocols, 8*, 265–281.

Butty, A. C., Perrinjaquet, N., Petit, A., Jaquenoud, M., Segall, J. E., Hofmann, K., Zwahlen, C., & Peter, M. (2002). A positive feedback loop stabilizes the guanine-nucleotide exchange factor Cdc24 at sites of polarization. *The EMBO Journal, 21*, 1565–1576.

Chant, J., & Herskowitz, I. (1991). Genetic control of bud site selection in yeast by a set of gene products that constitute a morphogenetic pathway. *Cell, 65*, 1203–1212.

Chant, J., Mischke, M., Mitchell, E., Herskowitz, I., & Pringle, J. R. (1995). Role of Bud3p in producing the axial budding pattern of yeast. *The Journal of Cell Biology, 129*, 767–778.

Chant, J., & Pringle, J. R. (1991). Budding and cell polarity in *Saccharomyces cerevisiae*. *Current Opinion in Genetics and Development, 1*, 342–350.

Chant, J., & Pringle, J. R. (1995). Patterns of bud-site selection in the yeast *Saccharomyces cerevisiae*. *The Journal of Cell Biology, 129*, 751–765.

Chong, Y. T., Koh, J. L., Friesen, H., Duffy, S. K., Cox, M. J., Moses, A., … Andrews, B. J. (2015). Yeast proteome dynamics from single cell imaging and automated analysis. *Cell, 161*, 1413–1424.

Chung, C. Y., Funamoto, S., & Firtel, R. A. (2001). Signaling pathways controlling cell polarity and chemotaxis. *Trends in Biochemical Sciences, 26*, 557–566.

Cole, K. C., McLaughlin, H. W., & Johnson, D. I. (2007). Use of bimolecular fluorescence complementation to study in vivo interactions between Cdc42p and Rdi1p of *Saccharomyces cerevisiae*. *Eukaryotic Cell, 6*, 378–387.

Costanzo, M., Baryshnikova, A., Bellay, J., Kim, Y., Spear, E. D., Sevier, C. S., … Boone, C. (2010). The genetic landscape of a cell. *Science, 327*, 425–431.

Das, A., Slaughter, B. D., Unruh, J. R., Bradford, W. D., Alexander, R., Rubinstein, B., & Li, R. (2012). Flippase-mediated phospholipid asymmetry promotes fast Cdc42 recycling in dynamic maintenance of cell polarity. *Nature Cell Biology, 14*, 304–310.

Day, R. N., & Davidson, M. W. (2009). The fluorescent protein palette: Tools for cellular imaging. *Chemical Society Reviews, 38*, 2887–2921.

Drubin, D. G., & Nelson, W. J. (1996). Origins of cell polarity. *Cell, 84*, 335–344.

Eitzen, G., Thorngren, N., & Wickner, W. (2001). Rho1p and Cdc42p act after Ypt7p to regulate vacuole docking. *The EMBO Journal, 20*, 5650–5656.

Etienne-Manneville, S. (2013). Microtubules in cell migration. *Annual Review of Cell and Developmental Biology, 29*, 471–499.

Evangelista, M., Blundell, K., Longtine, M. S., Chow, C. J., Adames, N., Pringle, J. R., ... Boone, C. (1997). Bni1p, a yeast formin linking cdc42p and the actin cytoskeleton during polarized morphogenesis. *Science, 276*, 118–122.

Freifelder, D. (1960). Bud position in *Saccharomyces cerevisiae*. *Journal of Bacteriology, 80*, 567–568.

Freisinger, T., Klunder, B., Johnson, J., Muller, N., Pichler, G., Beck, G., ... Wedlich-Soldner, R. (2013). Establishment of a robust single axis of cell polarity by coupling multiple positive feedback loops. *Nature Communications, 4*, 1807.

Fujita, A., Lord, M., Hiroko, T., Hiroko, F., Chen, T., Oka, C., ... Chant, J. (2004). Rax1, a protein required for the establishment of the bipolar budding pattern in yeast. *Gene, 327*, 161–169.

Galbraith, C. G., & Galbraith, J. A. (2011). Super-resolution microscopy at a glance. *Journal of Cell Science, 124*, 1607–1611.

Gao, J. T., Guimera, R., Li, H., Pinto, I. M., Sales-Pardo, M., Wai, S. C., ... Li, R. (2011). Modular coherence of protein dynamics in yeast cell polarity system. *Proceedings of the National Academy of Sciences of the United States of America, 108*, 7647–7652.

Gardner, J. M., & Jaspersen, S. L. (2014). Manipulating the yeast genome: Deletion, mutation, and tagging by PCR. *Methods in Molecular Biology, 1205*, 45–78.

Ghaemmaghami, S., Huh, W. K., Bower, K., Howson, R. W., Belle, A., Dephoure, N., ... Weissman, J. S. (2003). Global analysis of protein expression in yeast. *Nature, 425*, 737–741.

Giaever, G., Chu, A. M., Ni, L., Connelly, C., Riles, L., Veronneau, S., ... Johnston, M. (2002). Functional profiling of the *Saccharomyces cerevisiae* genome. *Nature, 418*, 387–391.

Giaever, G., & Nislow, C. (2014). The yeast deletion collection: A decade of functional genomics. *Genetics, 197*, 451–465.

Goffeau, A., Barrell, B. G., Bussey, H., Davis, R. W., Dujon, B., Feldmann, H., ... Oliver, S. G. (1996). Life with 6000 genes. *Science, 274*(546), 563–567.

Goley, E. D., Rodenbusch, S. E., Martin, A. C., & Welch, M. D. (2004). Critical conformational changes in the Arp2/3 complex are induced by nucleotide and nucleation promoting factor. *Molecular Cell, 16*, 269–279.

Gulli, M. P., Jaquenoud, M., Shimada, Y., Niederhauser, G., Wiget, P., & Peter, M. (2000). Phosphorylation of the Cdc42 exchange factor Cdc24 by the PAK-like kinase Cla4 may regulate polarized growth in yeast. *Molecular Cell, 6*, 1155–1167.

Haustein, E., & Schwille, P. (2007). Fluorescence correlation spectroscopy: Novel variations of an established technique. *Annual Review of Biophysics and Biomolecular Structure, 36*, 151–169.

Hicks, J. B., Strathern, J. N., & Herskowitz, I. (1977). Interconversion of yeast mating types III. Action of the homothallism (HO) gene in cells homozygous for the mating type locus. *Genetics, 85*, 395–405.

Hinnebusch, A. G., & Johnston, M. (2011). YeastBook: An encyclopedia of the reference eukaryotic cell. *Genetics, 189*, 683–684.

Hirschberg, K., Miller, C. M., Ellenberg, J., Presley, J. F., Siggia, E. D., Phair, R. D., & Lippincott-Schwartz, J. (1998). Kinetic analysis of secretory protein traffic and characterization of golgi to plasma membrane transport intermediates in living cells. *The Journal of Cell Biology, 143*, 1485–1503.

Hoffman, G. R., Nassar, N., & Cerione, R. A. (2000). Structure of the Rho family GTP-binding protein Cdc42 in complex with the multifunctional regulator RhoGDI. *Cell, 100*, 345–356.

Howell, A. S., & Lew, D. J. (2012). Morphogenesis and the cell cycle. *Genetics, 190*, 51–77.

Howell, A. S., Savage, N. S., Johnson, S. A., Bose, I., Wagner, A. W., Zyla, T. R., … Lew, D. J. (2009). Singularity in polarization: Rewiring yeast cells to make two buds. *Cell, 139*, 731–743.

Huang, B., Babcock, H., & Zhuang, X. (2010). Breaking the diffraction barrier: Super-resolution imaging of cells. *Cell, 143*, 1047–1058.

Huang, B., Bates, M., & Zhuang, X. (2009). Super-resolution fluorescence microscopy. *Annual Review of Biochemistry, 78*, 993–1016.

Huh, W. K., Falvo, J. V., Gerke, L. C., Carroll, A. S., Howson, R. W., Weissman, J. S., & O'Shea, E. K. (2003). Global analysis of protein localization in budding yeast. *Nature, 425*, 686–691.

Irazoqui, J. E., Gladfelter, A. S., & Lew, D. J. (2003). Scaffold-mediated symmetry breaking by Cdc42p. *Nature Cell Biology, 5*, 1062–1070.

Itoh, R. E., Kurokawa, K., Ohba, Y., Yoshizaki, H., Mochizuki, N., & Matsuda, M. (2002). Activation of Rac and Cdc42 video imaged by fluorescent resonance energy transfer-based single-molecule probes in the membrane of living cells. *Molecular and Cellular Biology, 22*, 6582–6591.

Jares-Erijman, E. A., & Jovin, T. M. (2006). Imaging molecular interactions in living cells by FRET microscopy. *Current Opinion in Chemical Biology, 10*, 409–416.

Jo, M. C., Liu, W., Gu, L., Dang, W., & Qin, L. (2015). High-throughput analysis of yeast replicative aging using a microfluidic system. *Proceedings of the National Academy of Sciences of the United States of America, 112*, 9364–9369.

Johnson, J. M., Jin, M., & Lew, D. J. (2011). Symmetry breaking and the establishment of cell polarity in budding yeast. *Current Opinion in Genetics and Development, 21*, 740–746.

Johnston, G. C., Prendergast, J. A., & Singer, R. A. (1991). The *Saccharomyces cerevisiae* MYO2 gene encodes an essential myosin for vectorial transport of vesicles. *The Journal of Cell Biology, 113*, 539–551.

Jose, M., Tollis, S., Nair, D., Sibarita, J. B., & McCusker, D. (2013). Robust polarity establishment occurs via an endocytosis-based cortical corralling mechanism. *The Journal of Cell Biology, 200*, 407–418.

Jost, A. P., & Weiner, O. D. (2015). Probing yeast polarity with acute, reversible, optogenetic inhibition of protein function. *ACS Synthetic Biology, 4*, 1077–1085.

Kang, P. J., Beven, L., Hariharan, S., & Park, H. O. (2010). The Rsr1/Bud1 GTPase interacts with itself and the Cdc42 GTPase during bud-site selection and polarity establishment in budding yeast. *Molecular Biology of the Cell, 21*, 3007–3016.

Katta, S. S., Chen, J., Gardner, J. M., Friederichs, J. M., Smith, S. E., Gogol, M., … Jaspersen, S. L. (2015). Sec66-dependent regulation of yeast spindle-pole body duplication through Pom152. *Genetics, 201*, 1479–1495.

Kim, S. A., Heinze, K. G., & Schwille, P. (2007). Fluorescence correlation spectroscopy in living cells. *Nature Methods, 4*, 963–973.

Klunder, B., Freisinger, T., Wedlich-Soldner, R., & Frey, E. (2013). GDI-mediated cell polarization in yeast provides precise spatial and temporal control of Cdc42 signaling. *PLoS Computational Biology, 9*, e1003396.

Koch, G., Tanaka, K., Masuda, T., Yamochi, W., Nonaka, H., & Takai, Y. (1997). Association of the Rho family small GTP-binding proteins with Rho GDP dissociation inhibitor (Rho GDI) in *Saccharomyces cerevisiae*. *Oncogene, 15*, 417–422.

Koh, J. L., Chong, Y. T., Friesen, H., Moses, A., Boone, C., Andrews, B. J., & Moffat, J. (2015). CYCLoPs: A comprehensive database constructed from automated analysis of protein abundance and subcellular localization patterns in *Saccharomyces cerevisiae*. *G3 (Bethesda), 5*, 1223–1232.

Kost, B. (2008). Spatial control of Rho (Rac-Rop) signaling in tip-growing plant cells. *Trends in Cell Biology, 18*, 119–127.

Kozminski, K. G., Beven, L., Angerman, E., Tong, A. H., Boone, C., & Park, H. O. (2003). Interaction between a Ras and a Rho GTPase couples selection of a growth site to the development of cell polarity in yeast. *Molecular Biology of the Cell, 14*, 4958–4970.

Kozubowski, L., Saito, K., Johnson, J. M., Howell, A. S., Zyla, T. R., & Lew, D. J. (2008). Symmetry-breaking polarization driven by a Cdc42p GEF-PAK complex. *Current Biology, 18*, 1719–1726.

Layton, A. T., Savage, N. S., Howell, A. S., Carroll, S. Y., Drubin, D. G., & Lew, D. J. (2011). Modeling vesicle traffic reveals unexpected consequences for Cdc42p-mediated polarity establishment. *Current Biology, 21*, 184–194.

Lee, S., Lim, W. A., & Thorn, K. S. (2013). Improved blue, green, and red fluorescent protein tagging vectors for *S. cerevisiae*. *PLoS One, 8*, e67902.

Lee, S. S., Avalos Vizcarra, I., Huberts, D. H., Lee, L. P., & Heinemann, M. (2012). Whole lifespan microscopic observation of budding yeast aging through a microfluidic dissection platform. *Proceedings of the National Academy of Sciences of the United States of America, 109*, 4916–4920.

Leung, B. O., & Chou, K. C. (2011). Review of super-resolution fluorescence microscopy for biology. *Applied Spectroscopy, 65*, 967–980.

Levskaya, A., Chevalier, A. A., Tabor, J. J., Simpson, Z. B., Lavery, L. A., Levy, M., … Voigt, C. A. (2005). Synthetic biology: Engineering *Escherichia coli* to see light. *Nature, 438*, 441–442.

Li, R., & Bowerman, B. (2010). Symmetry breaking in biology. *Cold Spring Harbor Perspectives in Biology, 2*, a003475.

Li, R., & Gundersen, G. G. (2008). Beyond polymer polarity: How the cytoskeleton builds a polarized cell. *Nature Reviews. Molecular Cell Biology, 9*, 860–873.

Lippincott-Schwartz, J., Roberts, T. H., & Hirschberg, K. (2000). Secretory protein trafficking and organelle dynamics in living cells. *Annual Review of Cell and Developmental Biology, 16*, 557–589.

Liu, P., Young, T. Z., & Acar, M. (2015). Yeast replicator: A high-throughput multiplexed microfluidics platform for automated measurements of single-cell aging. *Cell Reports, 13*, 634–644.

Longtine, M. S., McKenzie, A., 3rd, Demarini, D. J., Shah, N. G., Wach, A., Brachat, A., … Pringle, J. R. (1998). Additional modules for versatile and economical PCR-based gene deletion and modification in *Saccharomyces cerevisiae*. *Yeast, 14*, 953–961.

Loren, N., Hagman, J., Jonasson, J. K., Deschout, H., Bernin, D., Cella-Zanacchi, F., … Braeckmans, K. (2015). Fluorescence recovery after photobleaching in material and life sciences: Putting theory into practice. *Quarterly Reviews of Biophysics, 48*, 323–387.

Makushok, T., Alves, P., Huisman, S. M., Kijowski, A. R., & Brunner, D. (2016). Sterol-rich membrane domains define fission yeast cell polarity. *Cell, 165*, 1182–1196.

Manukyan, A., Abraham, L., Dungrawala, H., & Schneider, B. L. (2011). Synchronization of yeast. *Methods in Molecular Biology, 761*, 173–200.

Marco, E., Wedlich-Soldner, R., Li, R., Altschuler, S. J., & Wu, L. F. (2007). Endocytosis optimizes the dynamic localization of membrane proteins that regulate cortical polarity. *Cell, 129*, 411–422.

Martin, S. G., & Arkowitz, R. A. (2014). Cell polarization in budding and fission yeasts. *FEMS Microbiology Reviews, 38*, 228–253.

Masuda, T., Tanaka, K., Nonaka, H., Yamochi, W., Maeda, A., & Takai, Y. (1994). Molecular cloning and characterization of yeast rho GDP dissociation inhibitor. *The Journal of Biological Chemistry, 269*, 19713–19718.

Mc Intyre, J., Muller, E. G., Weitzer, S., Snydsman, B. E., Davis, T. N., & Uhlmann, F. (2007). In vivo analysis of cohesin architecture using FRET in the budding yeast *Saccharomyces cerevisiae*. *The EMBO Journal, 26*, 3783–3793.

Meyvis, T. K., De Smedt, S. C., Van Oostveldt, P., & Demeester, J. (1999). Fluorescence recovery after photobleaching: A versatile tool for mobility and interaction measurements in pharmaceutical research. *Pharmaceutical Research, 16*, 1153–1162.

O'Neill, P. R., Kalyanaraman, V., & Gautam, N. (2016). Subcellular optogenetic activation of Cdc42 controls local and distal signaling to drive immune cell migration. *Molecular Biology of the Cell, 27*, 1442–1450.

Orlando, K., Sun, X., Zhang, J., Lu, T., Yokomizo, L., Wang, P., & Guo, W. (2011). Exo-endocytic trafficking and the septin-based diffusion barrier are required for the maintenance of Cdc42p polarization during budding yeast asymmetric growth. *Molecular Biology of the Cell, 22*, 624–633.

Ozaki-Kuroda, K., Yamamoto, Y., Nohara, H., Kinoshita, M., Fujiwara, T., Irie, K., & Takai, Y. (2001). Dynamic localization and function of Bni1p at the sites of directed growth in *Saccharomyces cerevisiae*. *Molecular and Cellular Biology, 21*, 827–839.

Ozbudak, E. M., Becskei, A., & van Oudenaarden, A. (2005). A system of counteracting feedback loops regulates Cdc42p activity during spontaneous cell polarization. *Developmental Cell, 9*, 565–571.

Park, H. O., & Bi, E. (2007). Central roles of small GTPases in the development of cell polarity in yeast and beyond. *Microbiology and Molecular Biology Reviews, 71*, 48–96.

Park, H. O., Chant, J., & Herskowitz, I. (1993). BUD2 encodes a GTPase-activating protein for Bud1/Rsr1 necessary for proper bud-site selection in yeast. *Nature, 365*, 269–274.

Park, H. O., Kang, P. J., & Rachfal, A. W. (2002). Localization of the Rsr1/Bud1 GTPase involved in selection of a proper growth site in yeast. *The Journal of Biological Chemistry, 277*, 26721–26724.

Parri, M., & Chiarugi, P. (2010). Rac and Rho GTPases in cancer cell motility control. *Cell Communication and Signaling, 8*, 23.

Prendergast, F. G. (1999). Biophysics of the green fluorescent protein. *Methods in Cell Biology, 58*, 1–18.

Pruyne, D., & Bretscher, A. (2000a). Polarization of cell growth in yeast. *Journal of Cell Science, 113*(Pt 4), 571–585.

Pruyne, D., & Bretscher, A. (2000b). Polarization of cell growth in yeast. I. Establishment and maintenance of polarity states. *Journal of Cell Science, 113*(Pt 3), 365–375.

Pruyne, D. W., Schott, D. H., & Bretscher, A. (1998). Tropomyosin-containing actin cables direct the Myo2p-dependent polarized delivery of secretory vesicles in budding yeast. *The Journal of Cell Biology, 143*, 1931–1945.

Puig, O., Caspary, F., Rigaut, G., Rutz, B., Bouveret, E., Bragado-Nilsson, E., ... Seraphin, B. (2001). The tandem affinity purification (TAP) method: A general procedure of protein complex purification. *Methods, 24*, 218–229.

Richman, T. J., Toenjes, K. A., Morales, S. E., Cole, K. C., Wasserman, B. T., Taylor, C. M., ... Johnson, D. I. (2004). Analysis of cell-cycle specific localization of the Rdi1p RhoGDI and the structural determinants required for Cdc42p membrane localization and clustering at sites of polarized growth. *Current Genetics, 45*, 339–349.

Rigler, R., Foldes-Papp, Z., Meyer-Almes, F. J., Sammet, C., Volcker, M., & Schnetz, A. (1998). Fluorescence cross-correlation: A new concept for polymerase chain reaction. *Journal of Biotechnology, 63*, 97–109.

Roemer, T., Madden, K., Chang, J., & Snyder, M. (1996). Selection of axial growth sites in yeast requires Axl2p, a novel plasma membrane glycoprotein. *Genes and Development, 10*, 777–793.

Sagot, I., Klee, S. K., & Pellman, D. (2002). Yeast formins regulate cell polarity by controlling the assembly of actin cables. *Nature Cell Biology, 4*, 42–50.

Sanders, S. L., Gentzsch, M., Tanner, W., & Herskowitz, I. (1999). O-Glycosylation of Axl2/Bud10p by Pmt4p is required for its stability, localization, and function in daughter cells. *The Journal of Cell Biology, 145*, 1177–1188.

Savage, N. S., Layton, A. T., & Lew, D. J. (2012). Mechanistic mathematical model of polarity in yeast. *Molecular Biology of the Cell, 23*, 1998–2013.

Schermelleh, L., Heintzmann, R., & Leonhardt, H. (2010). A guide to super-resolution fluorescence microscopy. *The Journal of Cell Biology, 190*, 165–175.

Schott, D., Ho, J., Pruyne, D., & Bretscher, A. (1999). The COOH-terminal domain of Myo2p, a yeast myosin V, has a direct role in secretory vesicle targeting. *The Journal of Cell Biology, 147*, 791–808.

Schwille, P. (2001). Fluorescence correlation spectroscopy and its potential for intracellular applications. *Cell Biochemistry and Biophysics, 34*, 383–408.

Sekar, R. B., & Periasamy, A. (2003). Fluorescence resonance energy transfer (FRET) microscopy imaging of live cell protein localizations. *The Journal of Cell Biology, 160*, 629–633.

Shaner, N. C., Campbell, R. E., Steinbach, P. A., Giepmans, B. N., Palmer, A. E., & Tsien, R. Y. (2004). Improved monomeric red, orange and yellow fluorescent proteins derived from *Discosoma sp.* red fluorescent protein. *Nature Biotechnology, 22*, 1567–1572.

Shcherbo, D., Murphy, C. S., Ermakova, G. V., Solovieva, E. A., Chepurnykh, T. V., Shcheglov, A. S., ... Chudakov, D. M. (2009). Far-red fluorescent tags for protein imaging in living tissues. *The Biochemical Journal, 418*, 567–574.

Sheff, M. A., & Thorn, K. S. (2004). Optimized cassettes for fluorescent protein tagging in *Saccharomyces cerevisiae*. *Yeast, 21*, 661–670.

Sikorski, R. S., & Hieter, P. (1989). A system of shuttle vectors and yeast host strains designed for efficient manipulation of DNA in *Saccharomyces cerevisiae*. *Genetics, 122*, 19–27.

Sit, S. T., & Manser, E. (2011). Rho GTPases and their role in organizing the actin cytoskeleton. *Journal of Cell Science, 124*, 679–683.

Slaughter, B. D., Das, A., Schwartz, J. W., Rubinstein, B., & Li, R. (2009). Dual modes of cdc42 recycling fine-tune polarized morphogenesis. *Developmental Cell, 17*, 823–835.

Slaughter, B. D., & Li, R. (2010). Toward quantitative "in vivo biochemistry" with fluorescence fluctuation spectroscopy. *Molecular Biology of the Cell, 21*, 4306–4311.

Slaughter, B. D., Schwartz, J. W., & Li, R. (2007). Mapping dynamic protein interactions in MAP kinase signaling using live-cell fluorescence fluctuation spectroscopy and imaging. *Proceedings of the National Academy of Sciences of the United States of America, 104*, 20320–20325.

Slaughter, B. D., Smith, S. E., & Li, R. (2009). Symmetry breaking in the life cycle of the budding yeast. *Cold Spring Harbor Perspectives in Biology, 1*, a003384.

Slaughter, B. D., Unruh, J. R., Das, A., Smith, S. E., Rubinstein, B., & Li, R. (2013). Non-uniform membrane diffusion enables steady-state cell polarization via vesicular trafficking. *Nature Communications, 4*, 1380.

Slaughter, B. D., Unruh, J. R., & Li, R. (2011). Fluorescence fluctuation spectroscopy and imaging methods for examination of dynamic protein interactions in yeast. *Methods in Molecular Biology, 759*, 283–306.

Smith, S. E., Rubinstein, B., Mendes Pinto, I., Slaughter, B. D., Unruh, J. R., & Li, R. (2013). Independence of symmetry breaking on Bem1-mediated autocatalytic activation of Cdc42. *The Journal of Cell Biology, 202*, 1091–1106.

Takahashi, Y., Nishimura, J., Suzuki, A., Ishibashi, K., Kinjo, M., & Miyawaki, A. (2008). Cross-talk-free fluorescence cross-correlation spectroscopy by the switching method. *Cell Structure and Function, 33*, 143–150.

Tcheperegine, S. E., Gao, X. D., & Bi, E. (2005). Regulation of cell polarity by interactions of Msb3 and Msb4 with Cdc42 and polarisome components. *Molecular and Cellular Biology, 25*, 8567–8580.

Thompson, B. J. (2013). Cell polarity: Models and mechanisms from yeast, worms and flies. *Development, 140*, 13–21.

Thompson, N. L. (1991). *Fluorescence correlation spectroscopy. Topics in fluorescence spectroscopy*. New York: Plenum Press.

Tiedje, C., Sakwa, I., Just, U., & Hofken, T. (2008). The Rho GDI Rdi1 regulates Rho GTPases by distinct mechanisms. *Molecular Biology of the Cell, 19*, 2885–2896.

Tischer, D., & Weiner, O. D. (2014). Illuminating cell signalling with optogenetic tools. *Nature Reviews. Molecular Cell Biology, 15*, 551–558.

Toettcher, J. E., Weiner, O. D., & Lim, W. A. (2013). Using optogenetics to interrogate the dynamic control of signal transmission by the Ras/Erk module. *Cell, 155*, 1422–1434.

Van Aelst, L., & D'Souza-Schorey, C. (1997). Rho GTPases and signaling networks. *Genes and Development, 11*, 2295–2322.

Van Munster, E. B., Kremers, G. J., Adjobo-Hermans, M. J., & Gadella, T. W., Jr. (2005). Fluorescence resonance energy transfer (FRET) measurement by gradual acceptor photobleaching. *Journal of Microscopy, 218*, 253–262.

Wagih, O., Usaj, M., Baryshnikova, A., VanderSluis, B., Kuzmin, E., Costanzo, M., … Parts, L. (2013). SGAtools: One-stop analysis and visualization of array-based genetic interaction screens. *Nucleic Acids Research, 41*, W591–W596.

Wedlich-Soldner, R., Altschuler, S., Wu, L., & Li, R. (2003). Spontaneous cell polarization through actomyosin-based delivery of the Cdc42 GTPase. *Science, 299*, 1231–1235.

Wedlich-Soldner, R., Wai, S. C., Schmidt, T., & Li, R. (2004). Robust cell polarity is a dynamic state established by coupling transport and GTPase signaling. *The Journal of Cell Biology, 166*, 889–900.

Weitzman, M., & Hahn, K. M. (2014). Optogenetic approaches to cell migration and beyond. *Current Opinion in Cell Biology, 30*, 112–120.

Winter, D., Podtelejnikov, A. V., Mann, M., & Li, R. (1997). The complex containing actin-related proteins Arp2 and Arp3 is required for the motility and integrity of yeast actin patches. *Current Biology, 7*, 519–529.

Winzeler, E. A., Shoemaker, D. D., Astromoff, A., Liang, H., Anderson, K., Andre, B., ... Davis, R. W. (1999). Functional characterization of the *S. cerevisiae* genome by gene deletion and parallel analysis. *Science, 285*, 901–906.

Witte, H., & Bradke, F. (2008). The role of the cytoskeleton during neuronal polarization. *Current Opinion in Neurobiology, 18*, 479–487.

Wu, C. F., Savage, N. S., & Lew, D. J. (2013). Interaction between bud-site selection and polarity-establishment machineries in budding yeast. *Philosophical Transactions of the Royal Society of London. Series B, Biological Sciences, 368*, 20130006.

Xie, Z., Zhang, Y., Zou, K., Brandman, O., Luo, C., Ouyang, Q., & Li, H. (2012). Molecular phenotyping of aging in single yeast cells using a novel microfluidic device. *Aging Cell, 11*, 599–606.

Yang, F., Moss, L. G., & Phillips, G. N., Jr. (1996). The molecular structure of green fluorescent protein. *Nature Biotechnology, 14*, 1246–1251.

Yang, X., Jost, A. P., Weiner, O. D., & Tang, C. (2013). A light-inducible organelle-targeting system for dynamically activating and inactivating signaling in budding yeast. *Molecular Biology of the Cell, 24*, 2419–2430.

Yu, J. H., Crevenna, A. H., Bettenbuhl, M., Freisinger, T., & Wedlich-Soldner, R. (2011). Cortical actin dynamics driven by formins and myosin V. *Journal of Cell Science, 124*, 1533–1541.

Zhang, N., & Bilsland, E. (2011). Contributions of *Saccharomyces cerevisiae* to understanding mammalian gene function and therapy. *Methods in Molecular Biology, 759*, 501–523.

Zhang, Y., Luo, C., Zou, K., Xie, Z., Brandman, O., Ouyang, Q., & Li, H. (2012). Single cell analysis of yeast replicative aging using a new generation of microfluidic device. *PLoS One, 7*, e48275.

Ziman, M., O'Brien, J. M., Ouellette, L. A., Church, W. R., & Johnson, D. I. (1991). Mutational analysis of CDC42Sc, a *Saccharomyces cerevisiae* gene that encodes a putative GTP-binding protein involved in the control of cell polarity. *Molecular and Cellular Biology, 11*, 3537–3544.

CHAPTER

Precise tracking of the dynamics of multiple proteins in endocytic events

3

A. Picco, M. Kaksonen[1]

University of Geneva, Geneva, Switzerland

[1]*Corresponding author: E-mail: marko.kaksonen@unige.ch*

CHAPTER OUTLINE

Abstract

Endocytosis is a complex and dynamic process that involves dozens of different proteins to define the site of endocytosis, form a membrane invagination, and pinch off a membrane vesicle into the cytoplasm. Fluorescent light microscopy is a powerful tool to visualize the dynamic behaviors of the proteins taking part in the endocytic process. The resolution of light microscopy is, however, a serious limitation. Here, we detail a fluorescence microscope method that we have developed to visualize the dynamics of the clathrin-mediated endocytic protein machinery in yeast cells. This method is based on

Methods in Cell Biology, Volume 139, ISSN 0091-679X, http://dx.doi.org/10.1016/bs.mcb.2016.11.002

subpixel centroid tracking of endocytic proteins. For each endocytic protein, the centroid trajectories obtained from multiple endocytic events are used to compute an average trajectory that describes, at nanometer scale, the assembly and movement of the protein during endocytosis. The average trajectories of the different endocytic proteins are then aligned together in space and time to reconstruct how the different proteins behave relative to each other during the endocytic process.

INTRODUCTION

Proteins interact together in space and time and organize in transient macromolecular machineries that regulate biological processes, such as membrane transport, cell division, and migration. The understanding of these biological processes requires a detailed study of the molecular architecture of these complex machineries: how the proteins assemble relatively to each other and how they operate together to perform their tasks.

Fluorescent light microscopy is a powerful tool for unraveling the details of dynamic processes because it allows direct visualization, in living cells, of fluorescently tagged proteins in action. However, fluorescent microscopy is diffraction limited, and the fluorescent signal of two fluorescent objects cannot be resolved if they are separated by a distance smaller than the so-called Abbe's limit. With oil immersion objectives, this distance is about 200 nm, which is also the maximum size of the endocytic protein machinery in yeast. Fluorescent objects whose size is below the resolution limit are imaged as spots that are larger than the objects themselves. Therefore, the diffraction barrier limits the insights that can be obtained from fluorescence microscopy as the proteins, and many of the structures the proteins assemble in, are smaller than Abbe's limit.

Digital imaging has lead to development of quantitative approaches to light microscopy where the pixel intensities are a measure of the number of photons emitted by the fluorophore and hence are taken as a measure of the likelihood that the fluorophore is positioned in the region imaged by each pixel. A fluorescent object, imaged as a diffraction limited spot, can then be located very precisely, with a localization precision that depends on the number of photons recorded and on the background noise. Localization microscopy approaches have been used to visualize the molecular architecture of complex molecular machineries, like the kinetochore (Joglekar, Bloom, & Salmon, 2009; Wan et al., 2009), or the endocytic machinery during plasma membrane invagination (Berro & Pollard, 2014; Berro, Sirotkin, & Pollard, 2010; Picco, Mund, Ries, Nédélec, & Kaksonen, 2015; Sirotkin, Berro, Macmillan, Zhao, & Pollard, 2010).

Clathrin-mediated endocytosis is the main vesicular trafficking pathway transporting material into eukaryotic cells (McMahon & Boucrot, 2011). Formation of endocytic vesicles is a complex process involving dozens of proteins that define the site of endocytosis, recruit cargo, and then invaginate a small region of the plasma membrane that is eventually separated into a vesicle. In yeasts, the endocytic process is highly stereotypical (Berro & Pollard, 2014; Kaksonen,

Sun, & Drubin, 2003; Mooren, Galletta, & Cooper, 2012; Picco et al., 2015), the whole process lasts few minutes but the final invagination of the plasma membrane and scission of the vesicle occur in less than 10 s (Weinberg & Drubin, 2012). During this short time window many molecular processes happen: actin filaments polymerize at the site of endocytosis to drive membrane invagination, the membrane is shaped into a tubular invagination together with the proteins that coat it and finally the vesicle is pinched off from the plasma membrane invagination and released into the cytoplasm. The method that we describe here allows one to follow the morphogenesis of the endocytic machinery during the invagination of the plasma membrane and the scission of the vesicle.

1. RATIONALE

When a fluorescently tagged endocytic protein is imaged with a fluorescence microscope, endocytic events appear as diffraction limited spots on the yeast cell surface (Fig. 1). The centroid position of each of these fluorescent spots is a measure of the center of mass of the tagged proteins at that endocytic site. The fluorescence intensity of the spot reflects the number of tagged protein molecules assembled at that site. A trajectory, which is a collection of centroid positions and fluorescent intensities ordered in time, measures how the location and the number of the protein molecules changes over time. As endocytosis in yeast is a highly stereotypical process, multiple trajectories of the same protein, tracked in different endocytic events, can be averaged together to obtain a trajectory that records the

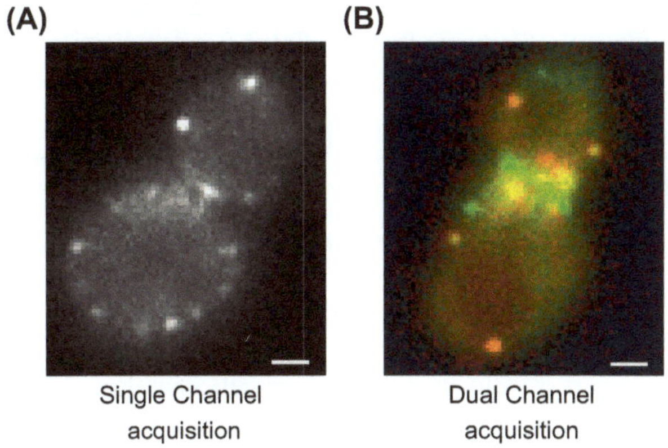

FIGURE 1

(A) Yeast cell expressing Sla1–GFP imaged at the equatorial plane of the cell. Sla1 localizes to the endocytic sites, which appear as diffraction limited spots on the edge of the cell. (B) Yeast cell expressing Sla1–GFP (green) and Abp1–mCherry (red) imaged at the equatorial plane of the cell. Both the green and red channels were imaged simultaneously. Scale bars are 1 μm.

average behavior of the protein at the endocytic site over time. The precision of the average trajectory is higher than the precision of the individual trajectories (Picco et al., 2015).

Here, we describe the procedure that we have used to image endocytic events in budding yeast *Saccharomyces cerevisiae*. With this method, we can track endocytic protein components, compute their average trajectories, and align, in space and time, the average trajectories of different endocytic proteins with a time precision below 1 s and a spatial precision of ~10 nm. The whole procedure can be divided in two main steps: In the first step, we compute the average trajectory of an endocytic protein tagged with green fluorescent protein (GFP). We will refer to this step as single channel (SC) acquisition as the imaging is done only in the channel of the tag of the protein of interest. If the goal is just to analyze the dynamic properties of individual endocytic proteins, the SC acquisition may be enough. In the second step, we align the average trajectories of different proteins together. We image simultaneously the protein of interest, tagged with GFP, together with another endocytic protein, which acts as a reference, tagged with the red fluorescent protein mCherry. The two proteins are tracked in each endocytic event, and their trajectory pair measures the relative position of the protein of interest in respect to the reference protein. These trajectory pairs are thus used to align in time and in space the average trajectory of the protein of interest to the average trajectory of the reference. If the reference protein is imaged together with different proteins of interest, the average trajectories of each of these proteins can be aligned all together (Picco et al., 2015). We will refer to this step as the dual channel (DC) acquisition, because the imaging is done on two channels simultaneously.

The two endocytic proteins that we use as an example here are Sla1, a coat associated protein, and Abp1, an actin binding protein, which is commonly used as a marker for the actin filaments that form at the endocytic site.

2. SAMPLE PREPARATION

Yeast cells are grown overnight on synthetic complete medium lacking tryptophan (SC-Trp) at 25°C on a shacking incubator. Before imaging, cells are diluted again in SC-Trp and are grown to log phase (OD600 = 0.6).

We image yeast cells on round 25 mm coverslips, (Menzel-Gläser, #1). For imaging, the coverslip is held in an Attofluor cell chamber (Molecular Probes), which is a circular metal holder. The sample preparation can be easily adapted to other types of sample holders. It is important that yeast cells adhere firmly to the coverslip, so that cell movement does not interfere with the tracking of endocytic events. To adhere yeast cells to the coverslip, we used the following procedure:

- Coat the center of the coverslip with a 40 μL drop of 1 mg/mL Concanavalin A (ConA). Incubate the coverslip with the ConA drop for 10 min at room temperature.

- Remove the ConA solution and wash carefully the coverslip 3 times with SC-Trp. Each time, wash by pipetting repeatedly 40 μL of SC-Trp on the central region of the coverslip that was incubated with ConA.
- In the center of the ConA coated coverslip, add a 40 μL drop of the yeast cells from the log-phase culture; incubate cells on the coverslip for 10 min to allow them to sediment and adhere on the coverslip.
- Finally, wash the center of the coverslip repeatedly with 40 μL of SC-Trp to remove the cells that are not fully adhered.
- For imaging, the coverslip is left with a 40 μL drop of SC-Trp. Cover the cell chamber with another coverslip to prevent the sample from drying.

A bead sample (Tetraspeck 0.1 μm microspheres from Invitrogen) is used to align the two channels in the DC acquisition, and it is prepared as follows:

- Vortex beads diluted in water to a density of $\sim 2.5 \times 10^7$ particles/μL for a minute at max speed to break down bead aggregates.
- Pipette 5 μL of beads on a coverslip and let the drop dry completely on the coverslip (~1 h at room temperature).
- The coverslip is then prepared as when imaging yeast cells, adding 40 μL of medium, to mimic the imaging conditions of the yeast samples.

3. FLUORESCENCE MICROSCOPY

3.1 MICROSCOPY SETUP

Any fluorescence microscope equipped with a suitable light source, filters, a high numerical aperture objective, and a sensitive camera can be used. We use an inverted wide-field epifluorescence microscope (Olympus IX81) equipped with a 488 nm laser for the excitation of the GFP and a 561 nm laser for the excitation of the mCherry. The emission light from the SC acquisition is filtered using the GFP-3035C-OMF single-band filter set (Semrock). For the DC acquisitions, we use an OBS-U-M2TIR 488/561 dichroic mirror (Semrock) and the Dual-View beam splitter (Optical Insights, LLC). For both SC and DC acquisitions, cells are imaged with a 100×/1.45NA TIRF objective (from Olympus), and the images are recorded with an EMCCD camera (Hamamatsu ImagEM EMCCD). In addition, an extra 1.6× magnification lens is used. Therefore, the cell is magnified 160 times and the camera pixels, which are 16 μm in size, correspond to 100 nm in the image. The microscope and the camera are controlled by MetaMorph software (Molecular Devices).

3.2 SIMULTANEOUS TWO COLOR ACQUISITION AND ALIGNMENT OF THE TWO CHANNELS

When imaging for the DC acquisition, the GFP and mCherry channels are split using the Dual-View beam splitter and are projected on the two halves of the camera sensor for simultaneous acquisition. As the trajectories obtained by the two channels are used to align together the average trajectories in space and in time, it is important that the two channels are acquired simultaneously so that there is no delay between the trajectories in each DC pair. The imaging of the two channels must be preceded by the alignment of the beam splitter and by the acquisition of images of Tetraspeck beads that are fluorescent in both channels. The bead images must be acquired using the same region of the camera sensor that will be used to image the cells. The alignment of the beam splitter and the imaging of the bead sample are performed in the beginning of each day of DC acquisitions.

Images of the Tetraspeck beads are used to compute the transformation that aligns the GFP and mCherry trajectories in each DC pair and corrects for the chromatic aberration and other possible warping effects. This transformation is applied to the DC trajectory pairs, before these trajectories are used to compute the alignment of the average trajectories together (see Section 6). To compute an accurate transformation, the beads need to be dense enough to sample homogeneously the field of view. At the same time, the beads need to be sparse enough so that the individual beads can be clearly resolved from each other for centroid determination. To get optimum bead density, we image relatively sparse beads (see Section 2), but we repeat the imaging 5–6 times, each time shifting the bead image by approximately 20–50 pixels in a random direction. Therefore, the bead images homogeneously cover the full field of view, and at the same time each bead spot could be easily tracked in the individual images (Fig. 2).

3.3 IMAGE ACQUISITION OF ENDOCYTIC EVENTS

The growth of the plasma membrane invagination during endocytosis is approximately orthogonal to the plasma membrane (Idrissi et al., 2008, 2012; Kukulski, Schorb, Kaksonen, & Briggs, 2012). Therefore, when yeast cells are imaged so that the plane of focus is on the cell's equatorial plane, the direction of the invagination growth can be assumed to be planar with the focal plane (Picco et al., 2015). The equatorial plane can be found by using bright field illumination to search for the focal plane where the cell appears largest. The endocytic events appear on the edge of the yeast cell as diffraction limited spots (Fig. 1). For many endocytic proteins the centroid of the spot moves inward as the membrane invagination forms and the vesicle is pinched off. Avoid small budded cells as the small buds are crowded with endocytic events that cannot be resolved from each other.

The SC movies used to compute the average trajectories need to be acquired at a high enough frame rate to record accurately the dynamics of the process that is being tracked. We are interested in tracking the dynamics of endocytic proteins during the vesicle budding process, which last about 10 s. Therefore, we acquire images with a frame rate of 8.5–12.5 Hz, collecting about 60–80 frames during the invagination of the plasma membrane.

The frame rate can be reduced when acquiring the DC images. In fact, in the trajectory pairs, it is the relative position of the trajectories that is important for

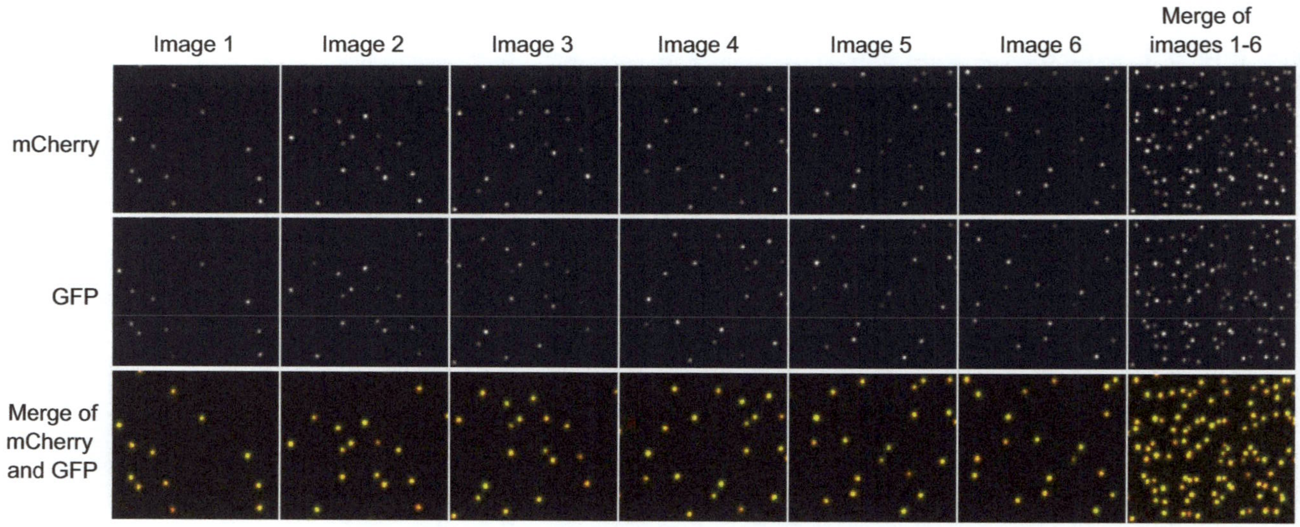

FIGURE 2

Beads are imaged in the two regions of the field of view, where the images of the GFP and mCherry channels are projected with an image splitter. Beads are sparse, so that they are easily resolvable. The bead images are repeatedly acquired on both channels moving the bead sample between each acquisition (Image 1–6). The series of bead images guarantees an even and dense distribution of the beads across the region of interest (Merge of images 1–6).

the alignment and not the fine details of their dynamics. In addition, the beam splitter reduces the image brightness and a slower frame rate allows longer exposure times that compensate for the loss of photons. However, one should note that excessively long exposure times tend to significantly smear fast dynamics. In the DC acquisitions, we use about half the frame rate of the SC acquisitions: 4–5 Hz.

Both the SC and DC movies should be long enough to record complete events and the excitation power should be adjusted to limit substantial photobleaching during the acquisition of the movie. In general, we acquire ~500–600 frames for the SC images and ~300 frames for DC images. In both SC and DC acquisition the sensitivity of the camera is optimized by adjusting the electron multiplying (EM) gain.

Finally, the images need few milliseconds to be read from the camera sensor. To minimize the delay in recording subsequent images, we cropped the field of view to record the signal from a region of the size of the cell, only. Importantly, this region should be kept constant through the acquisition of the different movies across the full experiment, so that the actual frame rates are conserved throughout all the acquisitions. The time interval between the frames, which is used later to time the trajectories, can be estimated as the average interval between the creation times of consecutive frames. The creation times can be obtained from the metadata of the movie in the image acquisition software. In a .tiff file the creation time information is labeled by the DateTime tag. On a computer running a Unix-based operating system (e.g., Mac OS X, Linux) the creation time of the frames can thus be obtained also by typing the following command in the Unix shell:

```
tiffinfo path/file_name.tiff | grep DateTime > datetime.txt
```

where `tiffinfo` is part of the Libtiff library (http://www.remotesensing.org/libtiff/), `path/file_name.tiff` are the directory path and the .tiff file name, and `datetime.txt` lists all the creation times of all the frames of the movie.

4. IMAGE PROCESSING

Before tracking the endocytic events, the SC and DC movies need to be corrected for general background, for photobleaching and for inhomogeneity in the local cellular background. The corrections are done using Fiji/ImageJ (version 1.50a or newer):

4.1 BACKGROUND SUBTRACTION

The background surrounding the cells can be corrected using the Rolling Ball algorithm:

- Select `Process > Subtract Background...` We use a `Rolling ball radius` of 90 pixels.

The radius should be as large as the yeast cell, the largest object in the image that is not part of the background.

4.2 PHOTOBLEACHING CORRECTION

The photobleaching correction is done by computing the mean fluorescence intensity of the whole cell in each frame. These mean intensities are then rescaled to be equal to the mean intensity measured in the first frame. First, one needs to select the image area that covers the whole cell, within which the average fluorescence intensity is measured:

- Select `Image > Adjust > Threshold...` and tick `Dark background`. If the selection of the yeast cell is not satisfactory adjust the threshold values accordingly;
- Select `Edit > Selection > Create Selection` to select the pixels within the cell outline.

Once the selection is created the `Threshold` dialog box can be reset. The selection of the cell outline allows you to measure the average pixel intensity within the cell in each frame of the movie. The average pixel intensity value of each frame needs to be rescaled to equal the average pixel intensity measured in the first frame, which is the least affected by photobleaching. Such operation can be done using the corr_bleach.txt macro available at http://www.embl.de/eamnet/html/bleach_correction.html.

4.3 LOCAL BACKGROUND CORRECTION

The fluorescence signal deriving from the labeled proteins at the endocytic site is overlaid with signal coming from the cytosolic pool of that protein and from the cellular autofluorescence. When yeast cells are imaged on their equatorial plane, the endocytic events appear at the edge of the cell image. As a consequence, the local fluorescence background at each endocytic spots is not homogeneous: at the cell edge the background signal forms a steep gradient that decreases when going from inside the cell toward the extracellular space. The inhomogeneous background shifts the centroid position along the gradient of the background fluorescence intensity. This effect is particularly evident when the average trajectories of different proteins with different cytoplasmic background intensities are aligned together. Therefore, the SC and both DC movies need to be corrected for the local background unevenness.

To correct for the inhomogeneous background we proceed as follows: We estimate the cell background by median filtering the cell in each frame. The radius of the kernel of the median filter should be as large as the diameter of the endocytic spots, so that they are replaced with the median intensity of the local background that surrounds them. It is important to note that while doing so, the median filter still preserves the cell boundaries. For images obtained with our microscope setup, we use a kernel radius of six pixels. The efficiency of the median filtering can be easily tested with the preview option in the median filter dialog box. The local background fluorescence is then corrected by subtracting the median filtered image from the original image, for each frame of the movie (Fig. 3). In summary, once the movie has been corrected for photobleaching:

- Select `Image > Duplicate...` and tick the option `Duplicate stack` to make a duplicate of the movie. This duplicate will be median filtered.
- Select the duplicated stack and choose `Process > Filters > Median...` to open the median filter dialog box. Enter the correct `Radius` (Radius = 6 pixels in our case).
- Select `Process > Image Calculator...`; `Image1` is the movie corrected for photobleaching, `Image2` is the median filtered movie and `Operation` is `Subtract`.

The resulting SC movie and the resulting pair of DC movies, corrected for local background inhomogeneity, will be used to track the dynamics of endocytic events and to align their average trajectories together.

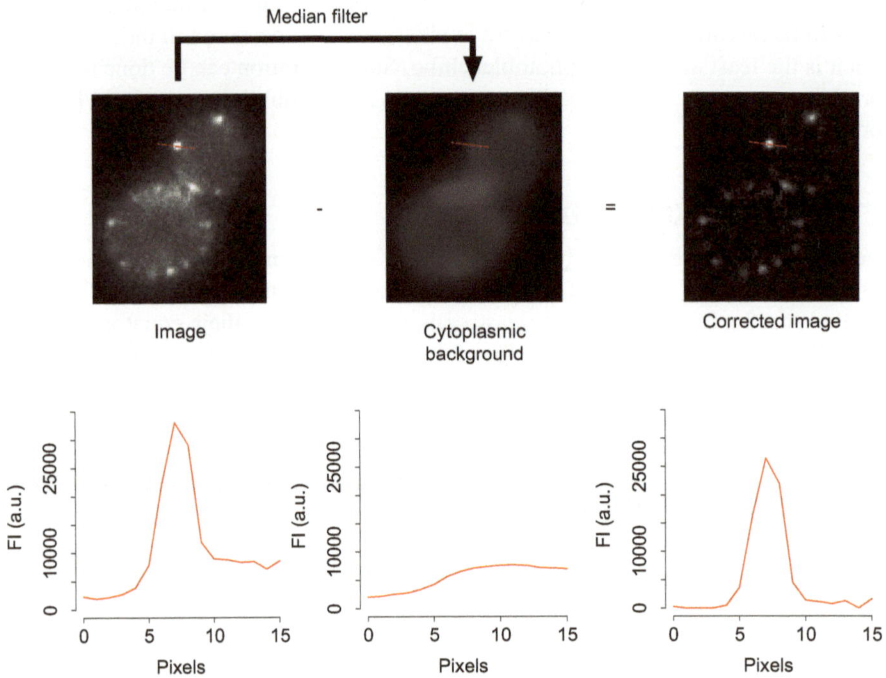

FIGURE 3

A median filter is used to compute an estimate of the cytoplasmic background in images already background subtracted and corrected for photobleaching. The median filtered image is subtracted from the image it is computed from. The plots show the fluorescence intensity profile along the *red dashed line* in the images.

5. TRACKING OF THE ENDOCYTIC SPOTS

The trajectory of each endocytic spot has a record of

- the different time points at which the spot position is measured. Usually, the frame indexing is used as a measure of the time;
- the x and y coordinates of the centroid position of the fluorescent spot in the image; and
- the fluorescence intensity of the spot, which is proportional to the number of labeled molecules located at that site.

The DC trajectory pairs are obtained by tracking the same endocytic event in both the GFP and mCherry channels.

For each trajectory, the time, the position coordinates, and the fluorescence intensity values should be recorded as separated columns in a .txt file. The choice of the tracking software and of the tracking algorithm is left to the reader. Averaging a large number of trajectories eventually will average out the difference in localization precision of different tracking algorithms. However, noisy trajectories can make it difficult to find the optimum alignment.

To track the endocytic spots we use ParticleTracker (http://imagej.net/Particle_Tracker, Sbalzarini & Koumoutsakos, 2005), which is a plug-in that runs under ImageJ/FIJI. In general, tracking software allows the user to set the parameters for the spot detection and tracking. We found that it is usually better to use a stringent set of parameters to avoid recording the noise in the images as false trajectories. In addition, the trajectories of endocytic events in crowded regions of the cell, such as the neck or the early bud, should be discarded as signal from the neighboring spots can affect tracking. Note that the sensitivity at which the endocytic spots are detected depends on the tracking parameters. The sensitivity will affect the observed lifetime of the spot (i.e., the number of frames where the spot is detected). More sensitive detection parameters will give longer lifetimes but will also lead to more false detections.

In ParticleTracker, we visualize the SC or DC trajectories of each endocytic event using the `Trajectories in Area Info` option. The coordinates of each individual trajectory are then copied and saved into different text files (e.g., `trajectory_01.txt`, `trajectory_02.txt`, etc.). All the trajectories tracked by ParticleTracker are also saved in a .txt file (e.g., `all_trajectories.txt`) by choosing the `Save full report` option. We manually curate the trajectory selection as the software occasionally fails to accurately track the endocytic spots: it can split a single trajectory in two or merge the trajectories of two neighboring endocytic events into one. We also manually check that the DC trajectories, each of which was tracked in its own image channel, are properly paired. To manually curate the trajectory selection we plot the selected trajectory (e.g., `trajectory_01.txt`) together with all the trajectories tracked by ParticleTracker (`all_trajectories.txt`) to confirm that the selected trajectory fully covers a single endocytic event.

To plot the selected trajectories we use Gnuplot (http://www.gnuplot.info). All trajectories are plotted in a 3-D plot, where the X and Y coordinates are defined by the centroid positions in the trajectory and the Z coordinate is used to plot the time. The trajectories are thus represented as a kymograph allowing us to easily assess the quality of each trajectory at each time point (Fig. 4). To plot the trajectories in Gnuplot, type:

```
set terminal x11
```

Or `wxt` instead of `x11` to be able to navigate the 3-D plot easily.

Then, type:

```
sp "path/all_trajectories.txt" u 2:3:1 w l
rep "path/trajectory_01.txt" u 2:3:1 w l
```

Where `path` is the directory path to the trajectory folder. The columns to be plotted on different axes are indicated by `2:3:1` (the second, the third, and the first column on the x-, y-, and z-axis, respectively). `all_trajectories.txt` is the .txt file where all trajectories tracked by ParticleTracker have been saved. In that file the

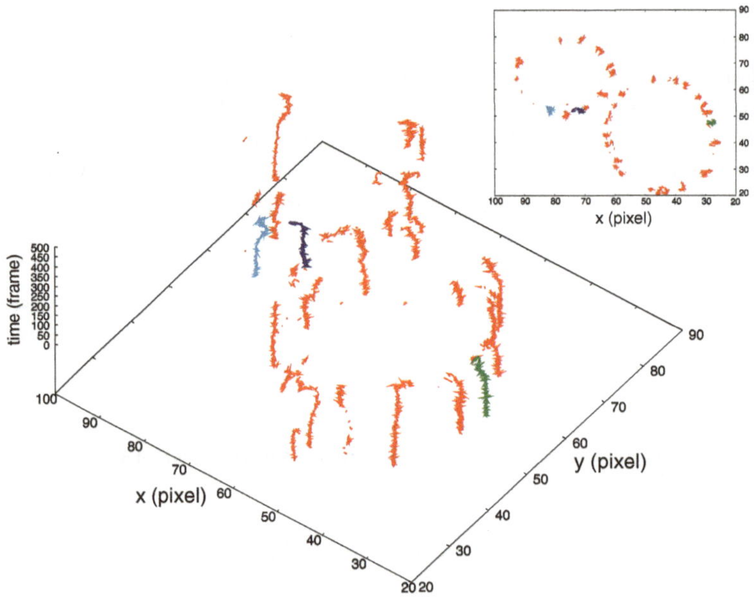

FIGURE 4

The dataset of the trajectories tracked by ParticleTracker are ploted as kymographs in Gnuplot (*red trajectories*). The *green, blue, and cyan trajectories* are the trajectories of three different endocytic spots that have been selected. These selected trajectories are overlaid on top of the whole ParticleTracker dataset. This allows the user to visually control that the selected trajectory describes the complete record of the selected endocytic spot and is not, for example, split into two separate trajectories. The x–y plot on the top right shows the trajectory view from the z-axis; the cell outline is recognizable.

first column lists the frame numbers while the second and third column list the trajectory coordinates. The different trajectories in `all_trajectories.txt` are separated by empty lines. `trajectory_01.txt` is one of the trajectories in `all_trajectories.txt` that tracks the endocytic event that has been selected.

The curated sets of SC or DC trajectories are then ready to be averaged or to be used to align together the average trajectories, respectively. An example of SC trajectories used to compute the average trajectory of Sla1 is shown in Fig. 5A. Fig. 5B shows an example of DC trajectory pairs used to align the average trajectory of Sla1 to the average trajectory of Abp1.

6. CORRECTION OF DUAL CHANNEL TRAJECTORY PAIR ALIGNMENT

To align the GFP and mCherry trajectories in each DC pair and to correct for the chromatic aberration and warping effects, we use the images of the bead samples. From both the GFP and mCherry images, we record the centroid positions of the beads with ParticleTracker, using the `Particle Detector only` option. The bead X and Y coordinates are saved in two columns and are grouped in two .txt files, one file for each channel (`beads_RFP.txt` and `beads_GFP.txt`). The bead centroid positions are then processed with Matlab (TheMathworks) and its Image Processing Toolbox to compute the warping transformation that needs to be applied to the DC trajectories. The bead positions measured in the two channels are not paired, and the number of beads found in the two channels can slightly differ because the differences in fluorescence intensities between the two channels can alter the bead recognition. We use the following Matlab script to pair the positions of the beads measured in the two channels and to discard the bead positions present only in one channel:

```
RFP = load( 'beads_RFP.txt' );
GFP = load( 'beads_GFP.txt' );

cutoff = 2;
paired_RFP = [];
paired_GFP = [];
l = length(RFP(:,1));

for i = 1:l
    d = sqrt( ( RFP(i,1) - GFP(:,1) ).^2 ...
            + ( RFP(i,2) - GFP(:,2) ).^2 );
    minima = d == min(d);
    if ( sum( minima ) == 1 && min(d) < cutoff )
        paired_RFP = [ paired_RFP ; RFP(i,:) ];
        paired_GFP = [ paired_GFP; GFP(find( minima ),:) ];
    end
end
```

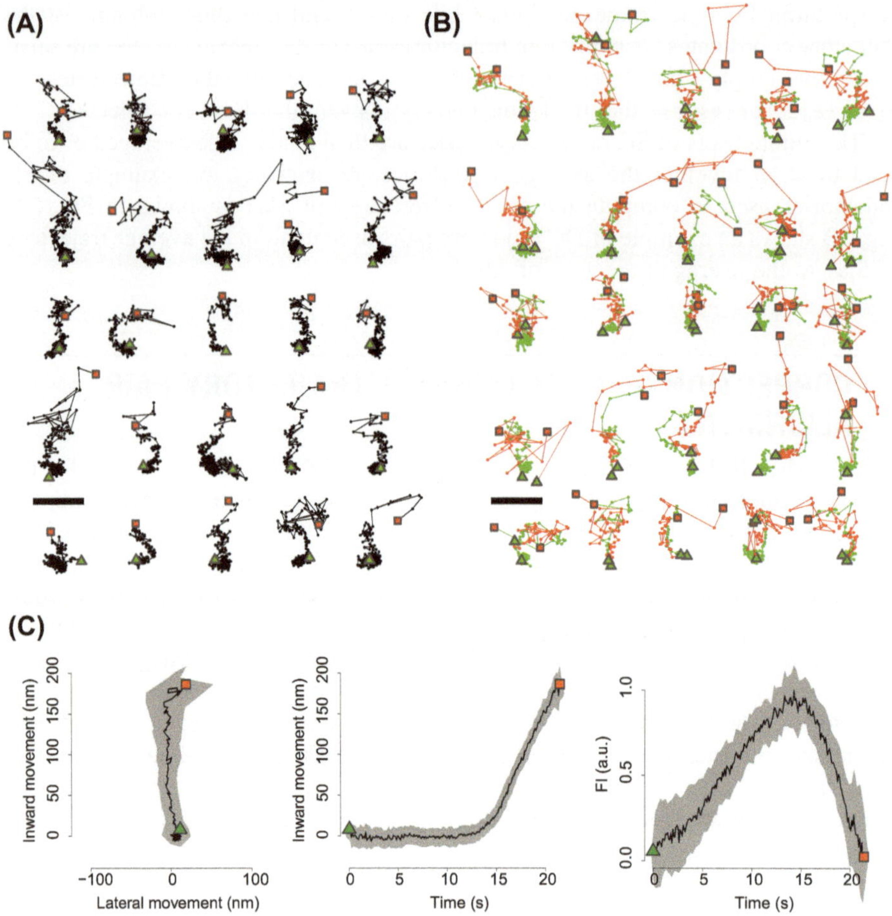

FIGURE 5

(A) Example of 25 single channel trajectories of Sla1–GFP, used to compute Sla1 average trajectory. (B) Example of 25 dual channel trajectory pairs of Sla1–GFP in green and Abp1–mCherry in red. These trajectories are used to align the average trajectory of Sla1 to the average trajectory of Abp1. (C) The average trajectory of Sla1–GFP. The left plot shows the lateral and inward movement of Sla1–GFP centroid. The central plot shows the inward movement over time. The right plot shows the change in the fluorescence intensity (FI) of Sla1 over time. The *shaded region* around the trajectories marks the 95% confidence interval. The *green triangle* marks the start of each trajectory, the *red square* marks its end.

The `cutoff` parameter is set to two pixels as the mCherry and GFP spots in our images are already well aligned by the beam splitter software, and it is not likely that their centroid positions differ for more than two pixels. However, this parameter could be different, depending on how the two channels are split and on the pixel size. The warping transformation that corrects for the misalignment and aberrations of the two channels is computed from the paired beads with the Image

Processing Toolbox of Matlab, using `cp2tform` with the local weighted mean `lwm` option and `N = 12` close control points:

```
warping = cp2tform( paired_GFP(:,1:2) , paired_RFP(:,1:2) , 'lwm' , N );
```

The transformation is then applied to one of the two channels of the pair of trajectories using the `tfrominv` function in Matlab:

```
trajectory_RFP = load( trajectory_RFP_path );
RFP_warped_coordinates = tforminv( warping , trajectory_RFP(:,2:3) );
trajectory_RFP_warped = [ trajectory_RFP(:,1) , RFP_warped_coordinates ,
trajectory_RFP(:,4:length( trajectory_RFP(1,:) )) ];
```

The `trajectory_RFP_warped` is saved to a .txt file and it supplants the original trajectory from the mCherry channel. The mCherry channel in each pair of trajectories is replaced with a warped trajectory. The new pairs of trajectories are those used to compute the alignment of the average trajectories together. For the details and updates about the computation of the warping transformation see Matlab help and references therein.

7. GENERATION OF AVERAGE TRAJECTORIES AND ALIGNMENT OF THE AVERAGE TRAJECTORIES

We developed a software program (Picco et al., 2015) that

- aligns in space and time individual SC trajectories to compute an average SC trajectory;
- aligns different average SC trajectories together, using the DC trajectories to compute their correct alignment.

The software runs in Python3.5 (www.python.org). The latest version of the software and its documentation can be downloaded from GitHub (http://apicco.github.io/trajectory_alignment/) or requested from the authors. The software comes with a trajectory class designed to format and handle trajectories as well as to store annotations about the trajectories and other experimental parameters. To compute the average trajectory, the software reads the SC trajectories, finds the optimal transformation that aligns them in time and in space, and, once all the trajectories are aligned, computes their average coordinate at each time point (Fig. 5C). The average is saved in a new text file. To align the average trajectory of a target protein to the average trajectory of Abp1, the software uses the relative position of the trajectories of the two proteins in each DC trajectory pair and outputs a new text file with the average trajectory of the target protein aligned, in time and in space, to Abp1 (Fig. 6A). You should refer to the software documentation for up-to-date instructions on its use and for examples.

Once the average trajectory of each protein is aligned to Abp1, it is possible to resolve the relative positions on the focal plane (Fig. 6A and B), the differences in the inward movement dynamics (Fig. 6B), the assembly and disassembly properties of each protein (Fig. 6C), and the relative positions of the protein centers of mass along the plasma membrane invagination (Fig. 6D).

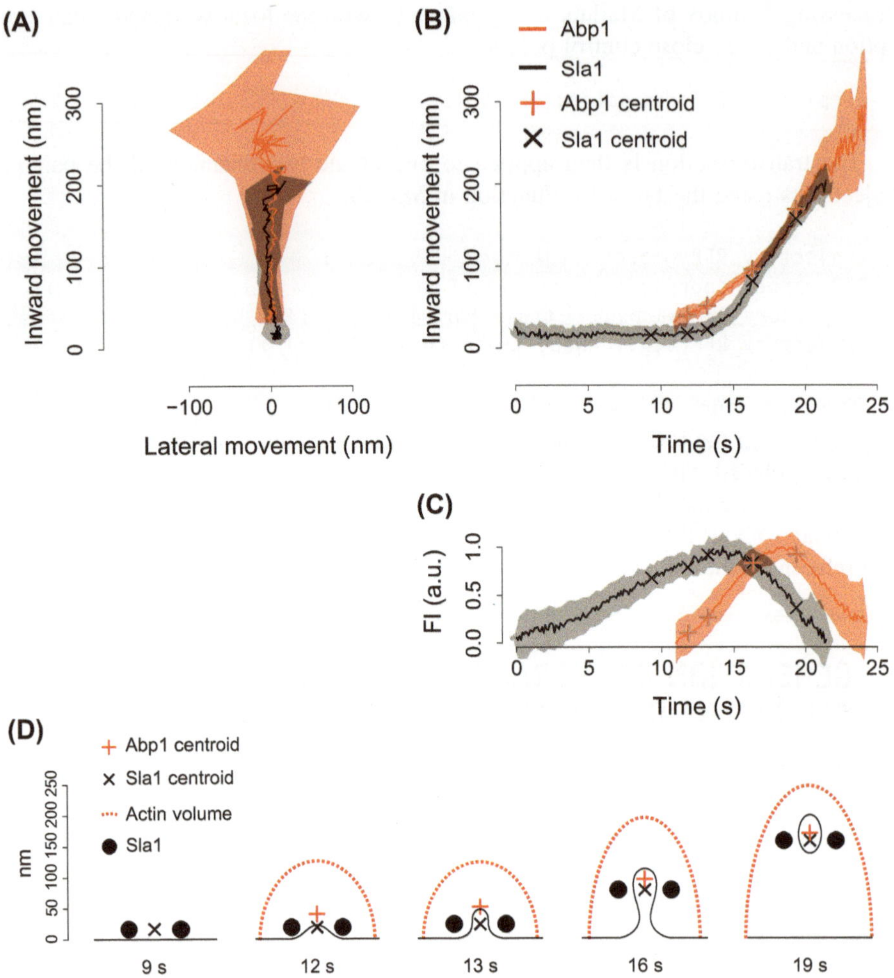

FIGURE 6

(A) The average trajectories of Sla1 (black) and Abp1 (red) are aligned together in space and time. (B) The inward movement of Sla1 and Abp1 average trajectories is plotted over time. (C) The fluorescence intensity (FI) profiles of Sla1 and Abp1 are aligned in time. The *shaded region* around the trajectories marks the 95% confidence interval. (D) The trajectories reveal the centers of mass of the proteins during the endocytic process. The center of mass position, marked by Abp1 and Sla1 average trajectories, can be aligned with the plasma membrane invagination (grey) and with the actin volume (*dashed red line*) profiles obtained from correlative light and electron microscopy (Kukulski et al., 2012; Picco et al., 2015). The *crosses* in (B) and (D) mark the same centroid positions; the *crosses* in (C) mark the FI at these centroid positions.

In general, 40–50 SC trajectories are sufficient to compute an average trajectory. For the alignment of the average trajectories, we acquire between 50 and 350 pairs of DC trajectories. The more DC trajectory pairs one can use the more accurate the alignment of the average trajectories will be.

The centroid position of many endocytic proteins moves into the cell as the plasma membrane is invaginated. However, for some proteins the centroid position remains stationary throughout all the endocytic process and it is not possible to estimate the direction of invagination from their trajectories. As the direction of invagination is the direction along which the SC trajectories are aligned in space, the SC trajectories of those proteins cannot be aligned. Nevertheless, their average trajectory can be computed using the DC trajectories only, where the Abp1 trajectories serve both as a reference and as an indicator for the direction of invagination. Such average is less accurate as the DC trajectories are in general noisier and the longer time interval between subsequent frames, needed to compensate for the noisy images, can smear dynamics, especially if the protein movement is so mild to challenge the alignment and averaging of the SC trajectories alone.

Ideally, the trajectories should cover the whole lifetime of the events. However, if the late vesicle budding stage is under study, it is also possible to average and align trajectories that lack their beginning. This is especially useful for proteins that are dim but have a long lifetime at the endocytic sites: these proteins are challenging to be imaged over their whole lifetime without bleaching their weak signals and one may therefore want to include also the incomplete events imaged at the beginning of the movies.

CONCLUSION

Here, we detailed a method to obtain data with high temporal and spatial resolution to describe the assembly–disassembly dynamics and the motile behaviors of endocytic proteins in living yeast cells (Fig. 6). This method can be used to characterize the behavior of novel proteins involved in endocytosis or to resolve subtle effects of mutations in the mutated protein or in the other endocytic proteins.

The trajectory data can be complemented with other types of data. For example, the fluorescence intensity data can be calibrated to obtain estimates of the absolute numbers of protein molecules assembled at endocytic sites at any time (Joglekar, Salmon, & Bloom, 2008; Lawrimore, Bloom, & Salmon, 2011; Picco et al., 2015; Sirotkin et al., 2010). The trajectories can be used together with local photobleaching to study the assembly dynamics of the target structures (Picco et al., 2015). In addition, the centroid trajectories of the endocytic proteins can also be aligned to time-resolved membrane shape data obtained recently by correlative light and electron microscopy (CLEM). This alignment of light and electron microscopy data puts the protein dynamics in the context of the membrane shape changes and thereby facilitates unraveling of the function of the different endocytic proteins (Kukulski et al., 2012; Picco et al., 2015).

REFERENCES

Berro, J., & Pollard, T. D. (2014). Local and global analysis of endocytic patch dynamics in fission yeast using a new "temporal superresolution" realignment method. *Molecular Biology of the Cell, 25*, 3501–3514.

Berro, J., Sirotkin, V., & Pollard, T. D. (2010). Mathematical modeling of endocytic actin patch kinetics in fission yeast: disassembly requires release of actin filament fragments. *Molecular Biology of the Cell, 21*, 2905–2915.

Idrissi, F.-Z. Z., Blasco, A., Espinal, A., & Geli, M. I. I. (2012). Ultrastructural dynamics of proteins involved in endocytic budding. *Proceedings of the National Academy of Sciences of the United States of America, 109*, E2587–E2594.

Idrissi, F.-Z. Z., Grötsch, H., Fernández-Golbano, I. M., Presciatto-Baschong, C., Riezman, H., & Geli, M.-I. I. (2008). Distinct acto/myosin-I structures associate with endocytic profiles at the plasma membrane. *Journal of Cell Biology, 180*, 1219–1232.

Joglekar, A. P., Bloom, K., & Salmon, E. D. (2009). In vivo protein architecture of the eukaryotic kinetochore with nanometer scale accuracy. *Current Biology, 19*, 694–699.

Joglekar, A. P., Salmon, E. D., & Bloom, K. S. (2008). Counting kinetochore protein numbers in budding yeast using genetically encoded fluorescent proteins. *Methods in Cell Biology, 85*, 127–151.

Kaksonen, M., Sun, Y., & Drubin, D. G. (2003). A pathway for association of receptors, adaptors, and actin during endocytic internalization. *Cell, 115*, 475–487.

Kukulski, W., Schorb, M., Kaksonen, M., & Briggs, J. (2012). Plasma membrane reshaping during endocytosis is revealed by time-resolved electron tomography. *Cell, 150*, 508–520.

Lawrimore, J., Bloom, K. S., & Salmon, E. D. (2011). Point centromeres contain more than a single centromere-specific Cse4 (CENP-A) nucleosome. *Journal of Cell Biology, 195*, 573–582.

McMahon, H. T., & Boucrot, E. (2011). Molecular mechanism and physiological functions of clathrin-mediated endocytosis. *Nature Reviews Molecular Cell Biology, 12*, 517–533.

Mooren, O. L., Galletta, B. J., & Cooper, J. A. (2012). Roles for actin assembly in endocytosis. *Annual Review of Biochemistry, 81*, 661–686.

Picco, A., Mund, M., Ries, J., Nédélec, F., & Kaksonen, M. (2015). Visualizing the functional architecture of the endocytic machinery. *eLife, 4*.

Sbalzarini, I. F., & Koumoutsakos, P. (2005). Feature point tracking and trajectory analysis for video imaging in cell biology. *Journal of Structural Biology, 151*, 182–195.

Sirotkin, V., Berro, J., Macmillan, K., Zhao, L., & Pollard, T. D. (2010). Quantitative analysis of the mechanism of endocytic actin patch assembly and disassembly in fission yeast. *Molecular Biology of the Cell, 21*, 2894–2904.

Wan, X., O'Quinn, R. P., Pierce, H. L., Joglekar, A. P., Gall, W. E., DeLuca, J. G., … Salmon, E. D. (2009). Protein architecture of the human kinetochore microtubule attachment site. *Cell, 137*, 672–684.

Weinberg, J., & Drubin, D. G. (2012). Clathrin-mediated endocytosis in budding yeast. *Trends in Cell Biology, 22*, 1–13.

CHAPTER

Quantitative approaches for the study of microtubule aster motion in large eggs

4

H. Tanimoto, N. Minc[1]

Institut Jacques Monod, CNRS, UMR 7592, Paris, France

[1]*Corresponding author: E-mail: nicolas.minc@ijm.fr*

CHAPTER OUTLINE

Abstract

The proper positioning of microtubule (MT) asters underlies fundamental processes such as nuclear centration, cell polarity, division positioning, and embryogenesis. In large eggs and early blastomeres, MT asters may exhibit long range motions with atypical speed and precision to target their functional position. The biophysical mechanisms regulating such motions remain however largely unknown. The centration of sperm asters in sea urchin embryos is a stereotypical example of such aster long range motion. In this chapter, we describe methods developed in this system to (1) quantify sperm aster 3-D motion with confocal microscopy and automated image analysis and (2) severe a portion of astral MTs with a UV laser. These methods may serve as a template to dissect the generic mechanisms of aster motion and force production in other embryos and cell types.

Methods in Cell Biology, Volume 139, ISSN 0091-679X, http://dx.doi.org/10.1016/bs.mcb.2016.12.003

INTRODUCTION

The spatiotemporal control of microtubule (MT) aster position in the cell is crucial to various biological contexts and has received extensive attention in the context of cell division-plane specification (Elric & Etienne-Manneville, 2014; Kimura & Kimura, 2014; Kiyomitsu, 2014; McNally, 2013; Minc & Piel, 2012). To precisely control daughter cell size and shape, cells may sense their shape to place the division plane accordingly. MT asters, which are star-shaped arrays of MTs radiating from centrosomes, may serve to convert cell shape information to division-plane positioning. In small cells such as adherent animal cells, metaphase asters around the spindle attach to cell cortex and interact with molecular motors such as dynein at the cortex to mediate spindle and consequent division-plane positions (Kiyomitsu & Cheeseman, 2013). In contrast, in large cells such as echinoderm, amphibian, and fish eggs and early blastomeres, metaphase asters are too small to reach the cell cortex (Mitchison et al., 2012). In these cells, interphase/telophase asters fill large portion of the cytoplasm and have been proposed to mediate division plane determination through dynein pulling forces produced along MTs in bulk cytoplasm (Minc, Burgess, & Chang, 2011; Tanimoto, Kimura, & Minc, 2016; Wuhr, Tan, Parker, Detrich, & Mitchison, 2010).

The functional positioning of MT asters relies on precise aster motion. While many studies have focused on the mechanisms that determine aster final position, little is known about how asters move inside cells. This question becomes particularly prominent in the context of eggs and early blastomeres given their large size and the rapidity of early embryonic cell cycles. A typical example, conserved in most species, is sperm aster centration. After fertilization, sperms bring centrioles inside eggs to nucleate a sperm aster that moves to the exact center of the large egg (Chambers, 1939; Hamaguchi & Hiramoto, 1980). These asters continuously grow in size and move at speeds of several microns per minute to reach the exact center of these large cells. The lack of information on sperm aster motion may be in part associated to difficulties of imaging motion in 3-D, in large cells, and on the limited set of quantitative methodologies to manipulate aster forces in space and time. Recently, we developed quantitative methods, based on 3-D microscopy and laser ablation to address the biophysical mechanisms driving sperm aster motion in sea urchin embryos (Tanimoto et al., 2016). The first method uses confocal microscopy and MATLAB-based automated image analysis to precisely track aster motion in 3-D, for normal spherical eggs and for eggs shaped in microchambers. This method serves to obtain motion characteristics such as travelling distance, speed, and velocity orientation. The second method uses a UV laser ablation setup to severe astral MTs in different positions and monitor the consequence on speeds and trajectories. Since the presented methods do not rely on any peculiar features of the organism, we emphasize they could be useful to study aster motion questions in other systems.

1. THREE-DIMENSIONAL TRACKING OF MICROTUBULE ASTERS IN SEA URCHIN EMBRYOS

This section describes a basic method to track aster centration in 3-D with high precision. The method includes cell handling, confocal microscopy, polydimethylsiloxane (PDMS) chamber operation, and image analysis using a custom-made code written in MATLAB.

1.1 DETECTING 3-D ASTER POSITION USING CONFOCAL MICROSCOPY

Materials

Sea urchin (*Paracentrotus lividus*), sterile water, artificial sea water (ASW; Instant Ocean; Reef Crystals), 80 μm Nitex mesh (Sefar), Hoechst 33342 (Sigma), glass-bottom dish (50 mm diameter, thickness no. 1.5; MatTek Corporation), protamine sulfate (Sigma), PDMS chamber (prepared as described in (Chang, Atilgan, Burgess, & Minc, 2014)), plasma cleaner (Harrick Plasma), air gun, spinning disk confocal microscopy (Nikon Ti-Eclipse combined with Yokogawa CSU-X1 spinning-head and Hamamatsu EM-CCD camera) equipped with a 40× objective (Nikon Plan Fluor, NA 1.3).

1.1.1 Sample preparation

a. Preparation of imaging dishes. Treat glass-bottom dishes with 1% protamine sulfate for 2 min. Rinse the dishes thoroughly with sterile water, and dry them under the hood. This treatment makes the dishes positively charged, on which eggs will adhere. The coated dishes can be stored in a clean box at room temperature for at least 1 month.

b. Obtain gametes from adult sea urchins by intracoelomic injection of 0.5 M KCl. Keep sperm at 4°C, and use them within ~1 week. Rinse the eggs twice using ASW, and store them at 16°C. Eggs should be used within the day of collection.

c. Remove the jelly coat of unfertilized eggs to facilitate cell adhesion by passing them three times though an 80-μm Nitex mesh. Alternatively, acidic treatment may also be used (1−2 min treatment with pH 5.5 ASW at 16°C, see (Cheers & Ettensohn, 2004) for details).

d. Place a 6 mL solution of dejellied eggs on a coated glass-bottom dish with adequate dilution in ASW. Add the DNA dye Hoechst 33342 at a final concentration of 1 μg/mL. If the Hoechst signal intensity of the male pronucleus is too weak, one can increase Hoechst concentration up to 5 μg/mL, which does not alter aster centration dynamics in our experimental conditions.

1.1.2 Imaging

Setup

The position of MT asters is determined by tracking the center of the male pronucleus bound to centrosomes which act as microtubule organizing centers. The male pronucleus is visualized with the fluorescent DNA dye Hoechst, and its position is measured by taking confocal time-lapse Z stacks (Fig. 1A). The interval

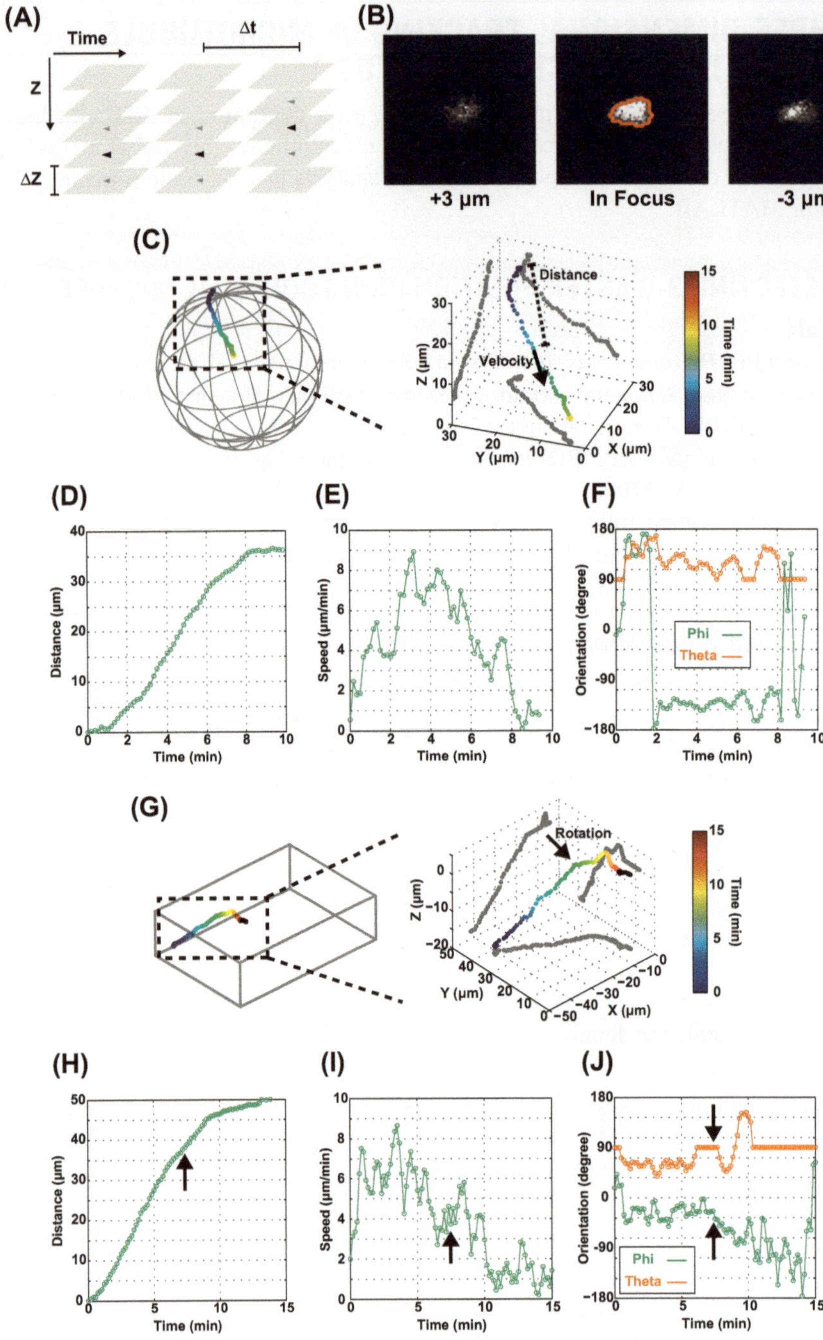

FIGURE 1 Three-dimensional tracking of centering microtubule (MT) asters.

of the Z slice should be small enough to precisely determine the position of the aster in Z. However, a too small slice interval reduces the temporal resolution and can cause phototoxicity. A Z-interval of 3 μm is sufficient to catch the male pronucleus (~2 μm size) in a single focal plane in one of the slice of the stack (Fig. 1B). The number of Z slices is set so that the confocal stack covers a hemisphere of the eggs, which allows to monitor the whole centration. Fifteen slices are sufficient to cover the hemisphere of a sea urchin egg (90 μm in diameter). The source laser is a 405 nm laser. Laser intensity and exposure time are adjusted to minimize phototoxicity. It takes ~8 s to take one Z stack in our setup, and the temporal interval Δt is set to 10 s.

Procedure

a. Place the sample on the microscope stage.
b. Dilute dry sperm 1000-fold in ASW by putting 4 μL sperm into 4 mL ASW. Activate them by vigorous aeration. Add the sperm solution dropwise to the dish under the microscope.
c. Detect the Z position of a sperm entering an egg. Choose the initial Z position of the time-lapse movie so that the Z stack volume includes both initial male pronucleus position and the egg center.
d. Start time-lapse acquisition. The entire aster centration takes ~10–15 min in normal conditions (Fig. 1C).

(A) Sperm pronuclei at the aster center are imaged at each time point by acquiring confocal z slices. The slice interval (ΔZ) is set to be 3 μm, and 15 slices are acquired so that the Z stack covers the hemisphere of a sea urchin embryo whose radius is 45 μm. Each Z stack takes ~8 s for the acquisition and the time interval (Δt) is set to be 10 s. (B) Confocal images of the sperm pronucleus visualized with a DNA dye Hoechst. Images in which the sperm pronucleus is in focus and 3 μm above/below are shown. The Z position of the center of the sperm aster is determined as the plane with maximum Hoechst signal. The XY position is determined as the center of mass of the 2-D shape of the nucleus extracted by thresholding (*red line*). (C–F) Quantification of aster 3-D centration. (C) 3-D trajectory of a centering MT aster in a normal spherical cell. $X = Y = Z = 0$ is the center of the egg (the same for (G)). The color indicates time. The trajectory consists from initial concaved penetration and long straight centration. The result shows that the precision of the tracking method is sufficient to confirm that the aster centration path is straight not only in XY but also in XZ and YZ. (D–F) Aster travelling distance (D), speed (E), and velocity orientation (F) are plotted as a function of time. See main text for definition. (G–J) Nonstraight aster centration in an egg shaped into a microfabricated chamber. The aster rotates during centration (indicated by a *black arrow*). Note that the tracking method identifies aster rotation in 3-D, which is characterized by changes in both polar and azimuthal angles (J).

1.1.3 Imaging in polydimethylsiloxane chamber

The 3-D tracking framework described in the last section can also be used to monitor aster motion in eggs shaped into PDMS chambers. This subsection describes additional procedures required for imaging aster motion in this condition.

Procedure

a. Prepare appropriate PDMS chambers with cell-size microwells (see (Chang et al., 2014) for detailed protocols for chamber design and fabrication).
b. Clean PDMS chambers and glass cover slips. The cover slips should be larger than the PDMS slabs. Treat them with acetone, isopropanol, and then rinse with sterile water. Dry them thoroughly using an air gun and keep them on kimwipes.
c. Activate PDMS chambers, with microwells facing up with a plasma cleaner. Check if they become hydrophilic by putting a drop of ASW. If not, dry them and repeat plasma activation possibly by increasing treatment time. Keep activated chambers in ASW until use. The activation lasts ~4 h.
d. Go to the microscope. Fertilize eggs in 15 mL falcon tubes. Wait ~1 min until the eggs accumulate at the bottom of the falcon tube by sedimentation. Place 40 μL of a dense egg solution on the clean coverslip.
e. Remove ASW from the activated PDMS slab with a kimwipe. Slowly apply the PDMS slab on the top of the egg solution with microwells facing down. Do not press the slab.
f. Remove excess ASW from the slab side using kimwipe. The slab approaches to the coverslip and causes eggs to go into the microwells.
g. Place the sample under the microscope. Find a cell successfully shaped in a well and acquire time-lapse images as described in Section 1.1.1 (Fig. 1G). It typically takes 3–5 min from sperm addition to the beginning of the image acquisition (steps d–g).

1.2 QUANTIFICATION OF ASTER 3-D MOTION

Materials

Standard desktop PC with a MATLAB software (supplemented with the statistical and image analysis toolboxes; MathWorks) and dedicated scripts (available upon demand).

Image analysis framework

The 3-D position of the male pronucleus is determined from a fluorescence time-lapse Z stack of images using home-built MATLAB scripts. Aster Z position is determined by detecting the plane in which the male pronucleus is in focus (Fig. 1A and B). Aster XY position is then determined by detecting the 2-D center of male pronucleus within the focal plane. The method enables a precise positional detection: the precision of the position in XY is estimated to be below 1 μm (using immobile sperm on the dish) and that in Z to be ~3 μm (set by the Z slice interval) (Fig. 1C).

The aster 3-D position $\overrightarrow{X}(t)$ is used to extract aster motion characteristics (Figs. 1D–F). Aster travelling distance $L(t)$ is defined as the distance between aster initial and current positions $L(t) = |\overrightarrow{X}(t) - \overrightarrow{X}(0)|$. Aster instantaneous velocity is defined as $\overrightarrow{V}(t) = \{\overrightarrow{X}(t+\tau) - \overrightarrow{X}(t)\}/\tau$, where $\tau = 30$ sec and $|V|$ is the aster speed. Aster 3-D orientation is defined using spherical coordinate;

$$\overrightarrow{V} = \begin{pmatrix} V_x \\ V_y \\ V_z \end{pmatrix} = \begin{pmatrix} |V| \sin\theta \cos\varphi \\ |V| \sin\theta \sin\varphi \\ |V| \cos\theta \end{pmatrix},$$

where θ and φ are polar and azimuthal angles, respectively.

Since aster motion involves fluctuations around a persistent directed motion, it may be nontrivial to define a travelling distance for asters undergoing a nonstraight motion. In the case of sperm asters' motion in shape-manipulated cells, aster rotations are very sharp so the migrating path consists of multiple straight lines and small part connecting them (Fig. 1G). The travelling distance of these asters can be defined as a summation of the passed paths and the distance from the last rotating position.

Procedure

a. Open MATLAB and run the script.
b. Choose the Z slice in which the male pronucleus is in focus at the first time point.
c. The script opens the image of the first time point. Select manually a rough position of the male pronucleus in the image.
d. The script updates the pronucleus position in the first time point by detecting the male pronucleus shape based on thresholding.
e. The scripts automatically find the Z plane in which the male pronucleus is in focus at successive time points and detect the XY position in the plane.
f. The script allows to check the tracking results in an interactive way. The script generates images of detected outline of male pronucleus superimposed on the real image of the male pronucleus at the focal plane. The tracking results are checked manually based on these images. If an error is found at a given time point, the aster position at this point may be updated by changing the initial position estimation and/or the threshold value for male pronucleus shape detection.

Output

The scripts generate an aster 3-D trajectory (Fig. 1C) and compute aster motion characteristics such as travelling distance (Fig. 1D) and amplitude and orientation of the instantaneous velocity (Figs. 1E and F). These outputs are saved both as plot figures and as MATLAB files which can be exported to excel. These quantifications can be used to characterize complicated aster motion such as rotation(s) in shape-manipulated cells (Fig. 1G–K).

2. LASER ABLATION OF ASTRAL MICROTUBULES AND IN SITU IMMUNOSTAINING

This section describes (1) a method to photoablate MTs on different parts of migrating sperm asters and (2) a method to perform rapid in situ immunostaining of ablated asters.

Materials

In addition to the materials required for tracking, the confocal microscopy is equipped with a high-power pulsed 355-nm UV laser system (iPulse, Roper Technologies) operated by iLas2 software (Roper Technologies). Chemicals for immunostaining, as listed in (Minc et al., 2011).

2.1 LASER ABLATION OF ASTRAL MICROTUBULES

Procedure

a. Prepare a sample and place it on the microscope as described in Section 1.1.1.
b. Fertilize eggs under the microscope by adding activated sperm.
c. Select an egg in which the sperm enters close to the equatorial plane so that the aster moves mostly in-plane.
d. Start a time-lapse acquisition to monitor the aster motion as described in Section 1.1.2. Since the aster moves almost in-plane, a Z stack is not always necessary for the measurement. We often reduce the number of Z slice to five and ΔZ to 1–2 μm.
e. Perform laser ablation when the aster has moved ~25 μm away from the cortex. The size of the target aster is a critical parameter to obtain a significant and reproducible effect of ablation on aster motion. The ablation position should be far enough from both the aster center and cell cortex. If the ablation position is too close to the male pronucleus, the laser irradiation bleaches the Hoechst signal and prevents subsequent tracking of the male pronucleus. In addition, ablating too close to the aster center oftentimes stops aster motion for unknown reason(s). If the ablation position is too close to the cell cortex, the irradiation affects cell integrity and oftentimes leads to cell bursting.
   We perform laser ablation when the aster has moved ~25 μm from the cortex, when the aster size is around 50 μm (Tanimoto et al., 2016). The irradiated line is set to be ~30 μm long. The irradiated position is set to be 10–15 μm away from the aster center (Fig. 2A). This setup enables us to perform laser ablation either at the front or on the side of MT asters. However, this setup does not work for rear MT ablation as the laser line tends to cross the cell cortex and yields to cell bursting typically 1 min after ablation.
   The laser power is set to be 50%. Each irradiated line consists of three 350-nm beam lines spaced by an interval of 1 μm (by setting the "line thickness" parameter in the iLas2 software to be two in the setting for the 40× objective). Each line irradiation takes 5–7 ms and is iterated 400 times. To ablate the MT aster in 3-D, the ablation is performed at three different heights: at the aster center and 5 μm, above and below. The total time needed to complete the

ablation is ~10 s, thus if the experiments do not need fine temporal resolution, ablations of several eggs may even be performed in parallel using a motorized XY stage. This setup is sufficient to significantly reduce MT length and density at the ablated region (Fig. 2B).

f. Continue time-lapse acquisition (Fig. 2C). Analyze aster motion as described in Section 1.2 to quantify the effects of ablation assay.

Laser ablation may sometimes cause small egg movement. In these cases, we determine the cell movement from the autofluorescence of the cytoplasm or differential interference contrast and subtract it from the aster motion to get the aster motion in the cell referential frame.

Output

The laser ablation assay of astral MTs serves to dissect the direction of MT forces that drives aster motion. In the case of sperm asters of sea urchin eggs, side ablation shifts the aster centration path away from the ablation side (Fig. 2D and E) and front ablation leads to transient reverse of aster velocity orientation (Tanimoto et al., 2016); both suggesting that aster motion is driven by pulling forces mediated by MTs.

2.2 IN SITU IMMUNOSTAINING OF ABLATED ASTERS

Procedure

a. Prepare PDMS slabs with a wide and straight perfusion channel and two inlet holes. The dimension of the channel is 8 mm * 25 mm and 70–75 μm in height. The height is ~10% smaller than the egg diameter to slightly flatten the eggs. The slight flattening stabilizes cell position and forces sperms to enter around the cell equatorial plane. Pierce holes at the two extremes of the channel to allow liquid perfusion.

b. Choose coverslips larger than the slab dimension. Clean the PDMS slabs as well as coverslips using acetone and isopropanol. Rinse them with water and thoroughly dry with an air gun.

c. Place a 25 μL solution of unfertilized egg onto the PDMS channel facing up. Put the coverslip onto the PDMS slab, and push onto the periphery to seal the channel.

d. Fertilize the eggs in the chamber by introducing activated sperm from the channel inlet hole.

e. Perform laser ablation as described in Section 2.1.

f. Perform fixation immediately after laser ablation, by rapidly introducing the fixative from the hole, following a fixation procedure described in (Minc et al., 2011). With this setup, one can fix samples within less than 10 s after ablation.

g. Perform successive steps of immunostaining by exchanging solutions under the microscope, to ensure that the eggs do not move nor deform during the process.

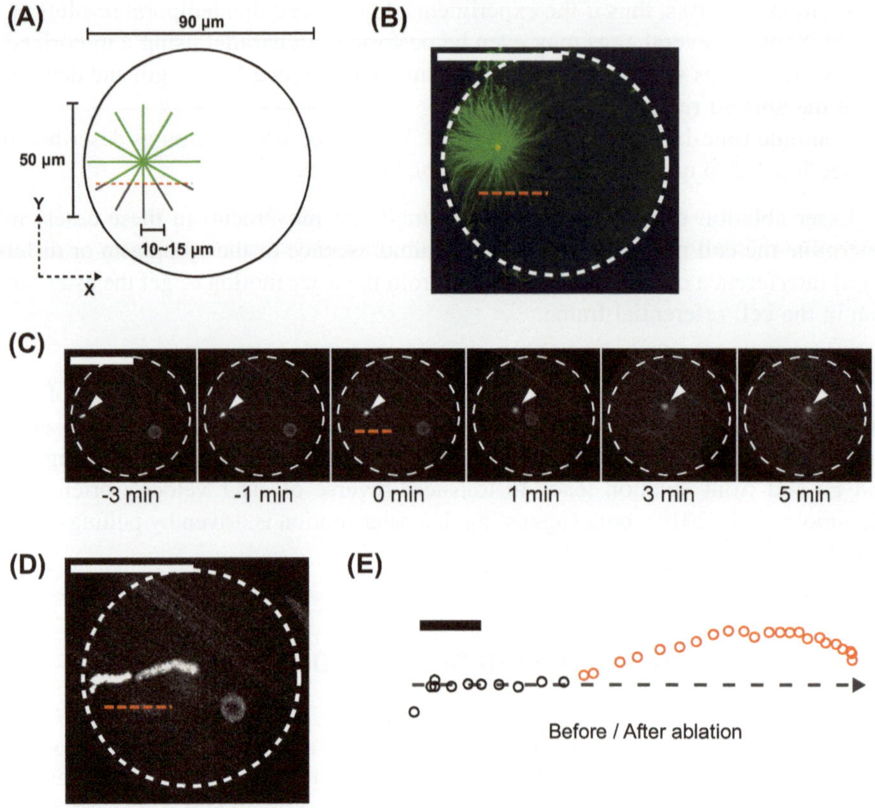

FIGURE 2 Laser ablation assay for microtubule (MT) asters.

(A) Schematic of the laser ablation assay. The setup for ablation at the aster side is shown. The *red broken line* indicates where the laser is applied (same for B–D). Intact and severed MTs are shown by green and gray colors, respectively. The ablation is performed in three Z heights: the aster focal plane and 5 μm above/below. (B) Tubulin immunostaining of an ablated aster. An aster was fixed in situ immediately after side laser ablation. Both density and length of MTs at the ablation side are significantly reduced. (C–E) Aster side ablation. (C) Time-lapse confocal images of the male pronucleus during aster side laser ablation. Time 0 corresponds to the moment when laser ablation is done. *Arrowhead*: male pronucleus. (D) Time projection of fluorescent images shown in (C). (E) The position of the ablated aster is shown every 10 s. The aster migrates from left to right, and *black and red colors* mark the aster center before and after ablation, respectively. The side ablation leads to an aster moving in a direction away from the MT ablation site, indicating the aster may be driven by outward forces. Bars: 50 μm (B–D) and 10 μm (E).

Output

Since the regrowth of ablated astral MTs may be very quick (Hyman, 1989), visualizing the effects of laser ablation requires rapid in situ immunostaining. The presented protocol enables rapid in situ immunostaining of ablated MT asters that is crucial to assess the effects of laser ablation on aster size and density (Fig. 2B).

CONCLUSIONS

The methods described in this chapter serve to study aster motion in sea urchin embryos and possibly for other species for which characterization of aster motion remains to be analyzed and understood (Stewart-Savage & Grey, 1982). The dynamics of aster motion reflects the underlying mechanisms of aster force production (Kimura & Onami, 2005, 2007). However, our studies suggest that MT asters in different model organisms could exhibit qualitatively different motions even if they may rely on common biological principles (Tanimoto et al., 2016). These observations should stimulate the developments of unified biophysical theories, which will likely require further quantitative experiments of aster motion in various cell types. Laser ablation is a powerful tool to dissect the direction of MT forces and may serve to obtain more quantitative information on the functional form of aster force dependency on aster size and shape; which are key for constructing quantitative mathematical models.

ACKNOWLEDGMENTS

We thank Akatsuki Kimura, François Rousset, and members of the Minc Laboratory for technical help and discussion. This work is supported by the CNRS and grants from the "Mairie de Paris emergence" program, the FRM "amorçage" grant AJE20130426890 and the ERC (CoG Forcaster N° 647073).

REFERENCES

Chambers, E. L. (1939). The movement of the egg nucleus in relation to the sperm aster in the echinoderm egg. *Journal of Experimental Biology, 16*(4), 409–424.

Chang, F., Atilgan, E., Burgess, D., & Minc, N. (2014). Manipulating cell shape by placing cells into micro-fabricated chambers. *Mitosis: Methods and Protocols*, 281–290.

Cheers, M. S., & Ettensohn, C. A. (2004). Rapid microinjection of fertilized eggs. *Methods in Cell Biology, 74*, 287–310.

Elric, J., & Etienne-Manneville, S. (2014). Centrosome positioning in polarized cells: common themes and variations. *Experimental Cell Research, 328*(2), 240–248.

Hamaguchi, M., & Hiramoto, Y. (1980). Fertilization process in the heart urchin, *Clypeaster japonicus* observed with a differential interference microscope. *Development, Growth & Differentiation, 22*(3), 517–530.

Hyman, A. A. (1989). Centrosome movement in the early divisions of *Caenorhabditis elegans*: a cortical site determining centrosome position. *Journal of Cell Biology, 109*(3), 1185–1193.

Kimura, K., & Kimura, A. (2014). A novel mechanism of microtubule length-dependent force to pull centrosomes toward the cell center. *Bioarchitecture, 1*(2), 74–79.

Kimura, A., & Onami, S. (2005). Computer simulations and image processing reveal length-dependent pulling force as the primary mechanism for *C. elegans* male pronuclear migration. *Developmental Cell, 8*(5), 765–775.

Kimura, A., & Onami, S. (2007). Local cortical pulling-force repression switches centrosomal centration and posterior displacement in *C. elegans*. *Journal of Cell Biology, 179*(7), 1347–1354.

Kiyomitsu, T. (2014). Mechanisms of daughter cell-size control during cell division. *Trends in Cell Biology*.

Kiyomitsu, T., & Cheeseman, I. M. (2013). Cortical dynein and asymmetric membrane elongation coordinately position the spindle in anaphase. *Cell, 154*(2), 391–402.

McNally, F. J. (2013). Mechanisms of spindle positioning. *Journal of Cell Biology, 200*(2), 131–140.

Minc, N., Burgess, D., & Chang, F. (2011). Influence of cell geometry on division-plane positioning. *Cell, 144*(3), 414–426.

Minc, N., & Piel, M. (2012). Predicting division plane position and orientation. *Trends in Cell Biology, 22*(4), 193–200.

Mitchison, T., Wühr, M., Nguyen, P., Ishihara, K., Groen, A., & Field, C. M. (2012). Growth, interaction, and positioning of microtubule asters in extremely large vertebrate embryo cells. *Cytoskeleton, 69*(10), 738–750.

Stewart-Savage, J., & Grey, R. D. (1982). The temporal and spatial relationships between cortical contraction, sperm trail formation, and pronuclear migration in fertilized Xenopus eggs. *Wilhelm Roux's Archives of Developmental Biology, 191*(4), 241–245.

Tanimoto, H., Kimura, A., & Minc, N. (2016). Shape–motion relationships of centering microtubule asters. *Journal of Cell Biology, 212*(7), 777–787.

Wuhr, M., Tan, E. S., Parker, S. K., Detrich, H. W., 3rd, & Mitchison, T. J. (2010). A model for cleavage plane determination in early amphibian and fish embryos. *Current Biology, 20*(22), 2040–2045.

CHAPTER

5

Dissection and characterization of microtubule bundles in the mitotic spindle using femtosecond laser ablation

R. Buđa*,[1,a], K. Vukušić*,[1,a], I.M. Tolić*

**Ruđer Bošković Institute, Zagreb, Croatia*

[1]*Corresponding authors: E-mail: rbudja@irb.hr; kvukusic@irb.hr*

CHAPTER OUTLINE

[a]These authors contributed equally to this work.

Methods in Cell Biology, Volume 139, ISSN 0091-679X, http://dx.doi.org/10.1016/bs.mcb.2016.11.007

Abstract

The mitotic spindle is a highly organized and dynamic structure required for segregation of the genetic material into two daughter cells. Although most of the individual players involved in building the spindle have been characterized in vitro, a general understanding of how all of the spindle players act together in vivo is still missing. Hence, in recent years, experiments have focused on introducing mechanical perturbations of the spindle on a micron scale, thereby providing insight into its function and organization, as well as into forces acting in the spindle. Among different types of mechanical perturbations, optical ones are more flexible, less invasive, and more precise than other approaches. In this chapter, we describe a detailed protocol for cutting the microtubule bundles in human cells using a near-infrared femtosecond laser. This type of laser microsurgery provides the ability to precisely sever a single microtubule bundle while preserving spindle integrity and dynamics. Furthermore, we describe quantitative measurements obtained from the response of a severed microtubule bundle to laser ablation, which reveal the structure and function of individual parts of the spindle, such as the bridging fiber connecting sister k-fibers. Finally, the method described here can be easily combined with other quantitative techniques to address the complexity of the spindle.

INTRODUCTION

Cell division is a basic process in all living organisms involving separation of the genetic material into the two daughter cells. To achieve faithful separation, mammalian cell needs a robust and dynamic structure. That structure is the mitotic spindle, a highly organized self-assembly of microtubules and associated proteins (Pavin & Tolić, 2016). Despite intense efforts in the field, a quantitative understanding of large-scale forces resulting from the collective behavior of individual microtubules forming bundles and proteins associated with them, is still lacking (Dumont & Mitchison, 2009). To move forward in the understanding of main forces acting in the mitotic spindle, experiments in recent years have been focused on introducing mechanical perturbations in mitotic spindle in vivo. The idea of all perturbation techniques is to perform controlled modifications of a selected structure and to observe the reaction of the cell to those alterations. That approach allows deduction of the forces acting on the selected structure in an unperturbed system.

In last decades, optical manipulation became the main tool for introducing perturbations on a very small scale (Magidson, Loncarek, Hergert, Rieder, & Khodjakov, 2007). This is due to many advantages of optical over nonoptical manipulation techniques including easy integration with standard microscopy systems, high spatial and temporal resolution, and finally, minimal interaction with the sample (Maghelli & Tolic-Norrelykke, 2010, 2011). As a result, many studies have been performed using optical techniques such as laser cutting in diverse model systems, including *Schizosaccharomyces pombe* (Guarino, Cojoc, Garcia-Ulloa, Tolic, & Kearsey, 2014; Khodjakov, La Terra, & Chang, 2004; Raabe, Vogel, Peychl, & Tolic-Norrelykke, 2009; Rumpf et al., 2010; Tolic-Norrelykke, Sacconi, Thon, & Pavone, 2004; Vogel, Pavin, Maghelli, Julicher, & Tolic-Norrelykke, 2009), *C.elegans* (Bringmann & Hyman, 2005; Yanik et al., 2004), *Dictyostelium* (Brito et al., 2005), *Drosophila* (Maiato, Khodjakov, & Rieder, 2005; Maiato, Rieder, & Khodjakov, 2004), mammals (Botvinick, Venugopalan, Shah, Liaw, & Berns, 2004; Cojoc et al., 2016; Colombelli, Reynaud, Rietdorf, Pepperkok, & Stelzer, 2005; Colombelli, Reynaud, & Stelzer, 2005; Elting, Hueschen, Udy, & Dumont, 2014; Goudarzi et al., 2012; Kajtez et al., 2016; Klingner et al., 2014; La Terra et al., 2005; Milas & Tolić, 2016; Sikirzhytski et al., 2014; Stiess et al., 2010), and plants (Reinhardt, Frenz, Mandel, & Kuhlemeier, 2005).

In this protocol, we use a manipulation technique based on laser ablation with near-infrared (NIR, 1030 nm) femtosecond laser pulses. NIR laser pulses enable specific ablation of a submicron-scale structures in vivo with minimal collateral damage (Steinmeyer et al., 2010; Vogel, Noack, Hüttman, & Paltauf, 2005). The use of high-speed electro-optic modulator unit (EOM) and high repetition rate laser provide the flexibility in optimizing laser-pulse repetition rates. Lower pulse repetition rates yield reduced heat accumulation in the specimen, and therefore less collateral damage (Steinmeyer et al., 2010; Vogel et al., 2005). The ablation with femtosecond pulse duration occurs through nonlinear multiphoton absorption (Vogel et al., 2005) and has been widely used to study the function of microtubules (Botvinick et al., 2004), mitochondria (Watanabe et al., 2004), and other organelles in cultured cells as well as in tissues and whole organisms (Shen et al., 2005; Stiess et al., 2010; Vogel et al., 2007).

Here we present a detailed protocol for performing femtosecond laser microsurgery on microtubule bundles in metaphase spindles combined with swept field confocal (SFC) microscopy, used for characterization of the bridging microtubule fibers and laser ablation in U2OS cells (Milas & Tolić, 2016). Bridging fibers were previously described as antiparallel microtubule bundles linking two sister k-fibers and balancing the forces on kinetochores in metaphase (Kajtez et al., 2016; Simunić & Tolić, 2016; Tolić & Pavin, 2016). We focused on metaphase because it is a dynamic steady state which means that despite large fluctuations and directed fluxes in both physical and chemical processes, average amount and position of spindle components is constant over time (Dumont & Mitchison, 2009). By applying this method, one structure of interest could be easily severed, without causing any significant collateral damage, in order to study the role of specific structures in the force balance of the mitotic spindle. The method described here can also be easily applied in studies of various submicron structures in different model systems.

1. SAMPLE PREPARATION

1.1 CELL CULTURE

The cell line used is human osteosarcoma U2OS stably expressing CENP-A-GFP (protein of kinetochore complex), mCherry-α-tubulin, and photoactivatable (PA)-GFP-tubulin, which was a gift from Marin Barišić and Helder Maiato (Institute for Molecular Cell Biology, University of Porto, Portugal). Cells were grown in flasks in Dulbecco's modified Eagle's medium (DMEM) (1 g/L D-glucose, L-glutamine, pyruvate) (Lonza, Basel, Switzerland) supplemented with 10% of heat-inactivated Fetal Bovine Serum (FBS) (Sigma–Aldrich, St Louis, MO, USA), 50 μg/mL geneticin (Life Technologies, Waltham, MA, USA), and penicillin/streptomycin solution (Lonza) to a final concentration of 100 I.U./mL penicillin and 100 μg/mL streptomycin. The cells were kept at 37°C and 5% CO_2 in a Galaxy 170s humidified incubator (Eppendorf, Hamburg, Germany). It is important to note that all the components added to the DMEM were sterilized by filtering.

1.2 SAMPLE PREPARATION FOR IN VIVO IMAGING

When cells reached 80% confluence, DMEM medium was removed from the flask and the cells were washed with 5 mL of 1% PBS. Subsequently, 1 mL of 1% Trypsin/EDTA (Biochrom AG, Berlin, Germany) was added and the cells were incubated at 37°C and 5% CO_2 in a humidified incubator (Eppendorf). After 5 min incubation, trypsin was blocked by adding 5 mL of DMEM medium. Cells were counted using the improved Neubauer chamber (BRAND GMBH + CO KG, Wertheim, Germany) and 4.5×10^5 cells were seeded and cultured in 2 mL DMEM medium with same supplements (see Section 1.1) at 37°C and 5% CO_2 on 35 mm glass coverslip uncoated dishes with 0.17 mm (#1.5 coverglass) glass thickness (MatTek Corporation, Ashland, MA, USA). After one-day growth, 1 h prior to imaging, the medium was replaced with Leibovit's (L-15) CO_2-independent medium (Life Technologies), supplemented with 10% FBS (Life Technologies), 100 I.U./mL penicillin, and 100 μg/mL streptomycin. Replacement is necessary because DMEM medium contains autofluorescent components (e.g., phenol red indicator dye) which increases the background signal while imaging the fluorescent components of interest. It is important to note that all the chemicals used for cell culture were preheated to 37°C.

2. IMAGING

2.1 MICROSCOPY EQUIPMENT

Inverted scanning confocal system coupled with laser ablation module (Fig. 1) was employed in this work. A detailed description of the system follows.

2.1.1 General microscope equipment

Opterra I (Bruker Nano Surfaces, Middleton, WI, USA) Inverted Confocal System with High Speed Camera: a compact multipoint confocal laser scanner designed to interface with standard microscopes. It is capable of variable scan speeds in excess of 100 frames per second (fps).

The system includes:

- Nikon TI-E Inverted Microscope (Nikon, Tokyo, Japan) configured for Opterra compatibility.
- User-selectable apertures (aperture plate) (Pinholes: 30, 45, and 60 μm; Slits: 22, 35, 50, and 75 μm); Multiple scan modes allow optimization of resolution, speed, and light delivery. Generally, pinholes provide better resolution, while slits allow for higher speed acquisitions.
- Opterra device control unit and galvanometer power supply.
- Scanner control embedded software.
- Helios three-line laser launch.
- 405/488/561/640 nm diode lasers, laser power: 100/150/50/100 mW.

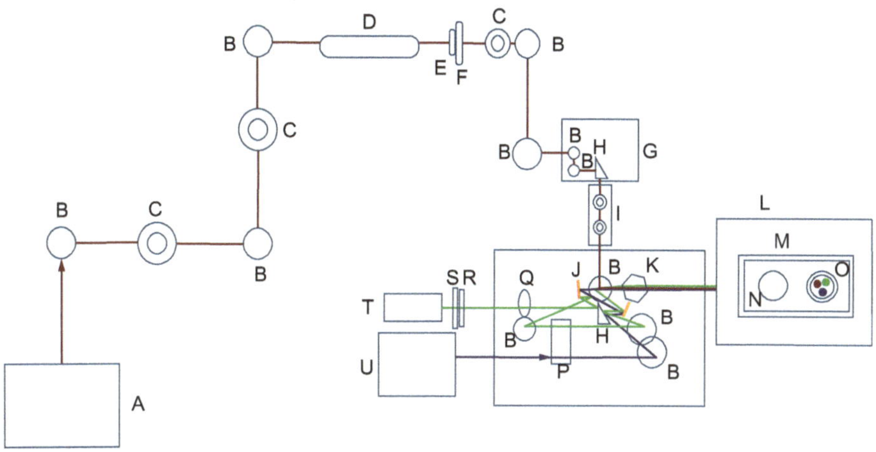

FIGURE 1 Detailed schematic representation of an imaging and laser ablation setup based on a near-infrared femtosecond-pulsed laser.

See Table 1 for the component list.

Table 1 Component List for the Optical System Layout of Fig. 1

	Description
A	Femtosecond laser
B	Kinematically mounted and internal mirrors (11)
C	Irises for alignment (3)
D	Electro-optic modulator (EOM)
E	Mechanical shutter
F	Half-wave plate
G	Laser coupler device
H	Galvanometer (2)
I	Tube with two lenses for alignment
J	Camera and microscope-side piezo
K	Dichroic mirror
L	Cage incubator system
M	x-y-z stage
N	Piezo z-stage plate
O	Nikon CFI Plan Apo VC 100 × Oil objective; NA 1.4 Oil
P	Aperture plate
Q	Emission filter wheel
R	2.0 × relay lens
S	Infrared (IR) light blocker
T	EMCCD camera
U	Excitation laser
Red line	Femtosecond laser beam path
Blue line	Excitation laser beam path
Green line	Emission beam path

- Opterra Dichroic & Barrier Filter Set, 405/488/561/640 nm (SPT); interchangeable sputter-coated pre-aligned and preset quadruple wavelength dichroic and barrier filter.
- Motorized 6-position filter wheel for emission wavelength selection.
- Epi-Fluorescence Light Source, X-Cite 120 for Nikon X-Cite120 lamp module (Excelitas Technologies, Waltham, MA, USA), 3 mm × 3000 mm liquid light guide, Nikon Microscope Collimating Adapter for standard epi-fluorescence illumination.
- Epi-Fluorescence Filter Set: B-2E/C FITC for Nikon.
- N-0965 Epi-Fluorescence Filter Set: G-2E/C TRITC for Nikon.
- Nikon CFI Plan Apo VC 20× Objective, NA0.75, WD 1.0 mm.
- Nikon CFI Plan Apo VC 100× Oil Objective; NA 1.4 Oil, WD 0.13 mm.
- Nikon CFI Plan APO Lambda 60× Oil Objective, NA 1.4 Oil WD 0.13 mm.

- ASI (Applied Scientific Instrumentation, Eugene, USA) X-Y-Z Stage with rotary encoders and piezo Z-stage plate (300 μm range with 4.5 nm resolution) with controller and adjustable slide holder.
- High-speed electron multiplying charge-coupled device (EMCCD) camera Evolve Delta 512 × 512 (Photometrics, Tucson, AZ, USA). EMCCD Digital Monochrome Camera with charged multiplication gain capability. FireWire interface. 20 MHz, Air Cooled.
- 2.0× camera relay lens (Bruker).
- High-Performance Imaging Workstation, custom-built with memory and speed required for Opterra systems.
- Prairie View 5.4 imaging software.
- Optical table with sufficient surface area (2.2 m × 1.5 m) and a set of four air-damped legs (TMC, Peabody, USA) with E 241/8/24 II compression unit (REM Machinen, Žeje, Slovenia).
- Uninterruptible power supply (UPS) unit VFI-2000TC LCD (Samurai power).

2.1.2 Laser ablation module

- Mikan air-cooled diode-pumped femtosecond oscillator (Amplitude Systèmes, Pessac, France)

  Main optical specifications:

- Wavelength: 1030 nm ± 10 nm.
- Average power: >1 W.
- Repetition rate: 50 MHz ± 5 MHz.
- Pulse duration: <250 fs (typically, 200 fs).
- Energy per pulse: 20 nJ.

  Beam conditioning optics:

- 302RM linear, high voltage, differential amplifier (E-O modulator) (302RM driver and 350-80 is the pockels cell) (ConOptics Inc., Danbury, CT, USA).
- Mechanical shutter system (Bruker) placed in the laser path after E-O modulator unit.
- Kinematically mounted mirrors (5) (Edmund Optics Inc., Barrington, IL, USA).
- Irises for alignment (Edmund Optics) (3).
- Half-wave plate and the rotation stage for 600−1050 nm high-power applications (Newport Corporation, Irvine, USA).
- Multiphoton shortpass emission filter 780−1064 nm (Chroma, Olching, Germany).
- Mounts (Thorlabs Inc, Newton, NJ, USA).

2.1.3 Sample maintenance equipment

- BoldLine Cage Incubator system H201 for maintaining constant temperature of sample environment and objectives (Okolab, Pozzuoli, NA, Italy).

2.2 IMAGING WORKFLOW FOR DIFFERENT APPLICATIONS

2.2.1 General protocol for spindle imaging

The steps involved in the experimental procedure are as follows:

1. Turn on the microscope system through the main switch followed by the computer, imaging lasers, camera, fluorescent lamp, and the femtosecond laser.
2. Optional for laser calibration procedure: place a custom-made calibration slide (microscopic slide with a permanent yellow fluorescent marker on it) in H301-K-frame with a microscopic slide insert. Choose Nikon CFI Plan Apo VC 100× Oil Objective and add appropriate immersion oil stable at 37°C (nd = 1515) (Biognost, Zagreb, Croatia). Using apochromatic lenses is especially important in experiments where co-localization of multiple fluorophores is performed because of their correction for spherical and chromatic aberrations (Balchand, Mann, & Wadsworth, 2016). Follow the laser calibration procedure in Prairie View 5.4. software. Be sure to optimize the value of laser power and exposure time depending on the type of the calibration slide to achieve ablation spots with clear edges. In our case, with a manually prepared calibration slide, the laser power was 50% and the laser exposure time was 300 ms. After performing the calibration procedure, save and load the most recent calibration file while performing further experiments. We recommend this procedure to be done at least once a month or when a laser misalignment during experiments is noticed.
3. About 30 min before imaging, turn on the temperature controller for cage incubator to raise the temperature of all components to 37°C. Place the temperature sensor in a petri dish in cage incubator H301-K-frame (Okolab) containing water for continuous reference temperature measurement.
4. Mount the petri dish with cells in cage incubator H301-K-frame (Okolab).
5. Choose Nikon CFI Plan Apo VC 60× Oil Objective and add appropriate immersion oil stable at 37°C (nd = 1515) (Biognost, Zagreb, Croatia).
6. In order to reduce photobleaching and photodamage prior to image acquisition, first observe cells with bright field optics.
7. Afterward, to observe tubulin marked with mCherry and kinetochores with GFP protein, visualize the cells using epifluorescence. Proper epifluorescent visualization is needed in order to find regular bipolar spindle in metaphase.
8. Use the settings A, described in further sections, for better insights into spindle structures and confirmation of normal spindle architecture. Exposure time and laser power must be adjusted so the fluorescent signal of interest can be detected.
9. Choose Nikon CFI Plan Apo VC 100× Oil Objective and add appropriate immersion oil stable at 37°C (nd = 1515) (Biognost, Zagreb, Croatia).
10. Prior to the start of the imaging adjust z-stacks for each cell using live mode with the z-piezo controller by defining top and bottom planes with an intention to encompass the central spindle planes.

11. Each individual cell must be positioned in the center of the scanned region to achieve maximal efficiency of laser ablation and to avoid cell movement out of the scanned region during imaging.
12. For imaging spindle structures in high-quality use settings B, described in further sections. For imaging the spindle dynamics use settings C, described in further sections. It is useful to observe the histogram of pixel intensity distribution for each fluorophore used to avoid saturated pixels in the imaging field.
13. After imaging, turn off all the parts of the system in the reversed order of turning the system on. Clean the used objective lenses using lens cleaning tissue (Whatman, Maidstone, UK) and 70% ethanol. Transfer the data from the imaging computer to a hard disk drive.

2.2.2 Settings for searching the cells (A)

After the cells were mounted onto the microscope stage, the 60× objective was chosen in order to get a bigger field of view (FOV) (68.3 × 68.3 μm) in which a greater number of cells in mitosis could be observed. To reduce photobleaching effects before imaging, the excitation 488 nm laser was on 10% laser power for GFP, the 561 nm laser for mCherry on 30% and 60 μm pinhole were used. Such power was sufficient to observe kinetochores aligned in the midzone of a bipolar metaphase spindle. Emissions from GFP and mCherry were detected in 498–558 and 585–665 nm ranges, respectively. Also, it was important to check that cells used in experiments express all fluorescent markers of interest. To further reduce the photobleaching effects, the imaging with EMCCD sensor camera was performed with 100 ms continuous exposure time and without any frame averaging.

2.2.3 Settings for imaging spindle structures in high quality (B)

In order to obtain the optimal balance between spatial resolution and signal-to-noise ratio, 60 μm pinhole aperture was used. The excitation 488 nm laser was on 40% laser power for GFP, and the 561 nm laser on 90% for mCherry. Emissions from GFP and mCherry were detected as previously mentioned. For high-quality imaging of all spindle components, 8–10 focal planes were imaged with 0.5 μm Z spacing using 2* frame averaging in order to get a better signal-to-noise ratio. The z-scan mode was unidirectional. The movement in the focal plane can be overcome by holding a set focus using a far red laser beam reflected off the surface of the coverslip. However, the use of that option is incompatible with NIR-femtosecond laser ablation because the laser beam cannot reach the sample. Imaging with EMCCD sensor camera was performed with 300 ms continuous exposure time. Electron multiplying gain was set at 600. Camera readout mode was 20 MHz. No binning was performed. Z device was Piezo with a step size of 0.5 μm. Final x, y pixel size in the image was 82 nm. Field of view during imaging was fixed at 41 × 41 μm. For ablation experiments (see Section 3), image acquisition was performed for 30–40 time frames at 20 s intervals. Even though the duration of one z-stack imaging was less than 20 s, the constant interval was chosen in order to reduce the loss of fluorescence intensity.

2.2.4 Settings for imaging the spindle dynamics (C)

In order to obtain an optimal balance between spatial resolution and signal-to-noise ratio, 60 μm pinhole aperture was used. The excitation 488 nm laser was on 30% laser power for GFP, and the 561 nm laser on 80% for mCherry. Emissions from GFP and mCherry were detected as mentioned above. Imaging with EMCCD sensor camera was performed with 500 ms continuous exposure time. Electron multiplying gain was set at 600. Camera readout mode was 20 MHz. No binning was performed. Z controller device was Piezo with a step size of 0.5 μm. In order to achieve faster acquisition, frame averaging was turned off and the fastest acquisition mode was turned on to take advantage of the higher speed acquisition capabilities of the piezo controller. Also, in this application, a region of interest (ROI) was used enabling us to image a smaller region (25 μm × 25 μm) encompassing just the mitotic spindle. That was done to obtain faster acquisition while the spindle was likely to maintain its position in this short period of imaging (1−3 min). Final x, y pixel size in the image was the same as aforesaid, 82 nm. For ablation experiments (see Section 3), image acquisition was performed for 10−20 time frames at 5 s intervals in order to capture the spindle dynamics without losing fluorescence intensity. To visualize the central part of the spindle and to prevent loss of the microtubule bundle from the imaged stack due to its movement in the z-direction after ablation, five focal planes were imaged with 0.5 μm Z spacing using unidirectional z-scan mode.

3. LASER ABLATION

The laser beam (1030 nm) for ablation was coupled to the photoactivation module of the microscope (Fig. 1). The setup for laser ablation was adjusted by choosing the Mark points—Live click and drag option in Prairie View Software 5.4. It is important to note that laser ablation was performed during live imaging. This option was crucial for our experiments because the spindle is a highly dynamic structure even at the metaphase steady state. Alternatively, ablation can be performed by manually choosing the desired shape, position, and size of the region for ablation before starting the imaging. In that way, the flexibility in choosing the exact z position of the cut is reduced compared to ablation during live imaging. Also, when not using the live mode, there is a greater possibility for unsuccessful ablation of the highly mobile structure. Different laser powers were tried and we decided to go with 70% of laser power, which corresponds to about 700 mW power at the sample plane. This power was kept constant during experiments because the very precise cuts with minimal side effects for cells were observed. The passage of the laser beam to the sample plane was controlled by the mechanical shutter placed in the laser path (Fig. 1). The shutter was coupled to open automatically with the ablation module. Hence, when the shutter was opened, the passage of the laser beam to the sample plane greatly increased the background noise, significantly reducing the signal-to-noise ratio. That disadvantage was solved by placing the IR emission filter in front of the camera.

Laser ablation could be performed in two modes, which provide high flexibility depending on the specific experimental application:

3.1 SINGLE-POINT MODE

The first mode (blank/unblank mouse cursor option) was the single-point mode with the fixed size of a region for ablation at 500 nm (minimal region for ablation) while the laser exposure time was controlled by holding the mouse button. The advantage of the single-point mode was the high precision combined with the flexible duration. However, the main disadvantage of the single-point mode was the lack of flexibility in controlling the size of the region for ablation.

3.2 SPIRAL MODE

In the spiral mode (pulse laser at mouse click option), the laser exposure time was fixed at 800 ms while the size of the region for ablation was varied between 500 nm and 1000 nm depending on the size of the spindle. The spiral mode means the laser is ablating multiple points along the shape of the spiral where the number of the spiral revolutions is adjustable. The laser exposure time at 800 ms was optimal for performing laser ablation in our system with minimal side effects. However, the duration of a laser beam could be adjusted depending on the specific application. It should be noted that when using this mode, the laser exposure time should be slightly increased to compensate for slow shutter opening (several 10 s of ms). Since the spiral mode was more flexible and more successful cuts in spindles of various sizes were observed, it was our predominant way of laser ablation.

3.3 PERFORMING LASER ABLATION

Prior to laser ablation, the general protocol for spindle imaging was applied (see Section 2.2.1.). Laser ablation was performed in live mode during the second time frame in z plane(s) in which a single microtubule bundle connecting a spindle pole with a kinetochore (k-fiber) could be discerned. Note that the imaging of one z plane could be shorter than the laser exposure time, which means that the ablation could encompass several z planes. The cut was performed on one of the outermost randomly chosen k-fibers, in one set of the experiments 2–2.5 μm and in the other 0.5–1.5 μm, from the kinetochore. In spite of the spindle movements in x, y, and z, ablating while continuously imaging enabled us to perform fast and precise cuts. Due to the fast reattachment of the ablated k-fibers to neighboring microtubules in U2OS cells (see Section 4.1.), multiple sequential cuts were sometimes necessary. When successful ablation was achieved, the imaging was continued.

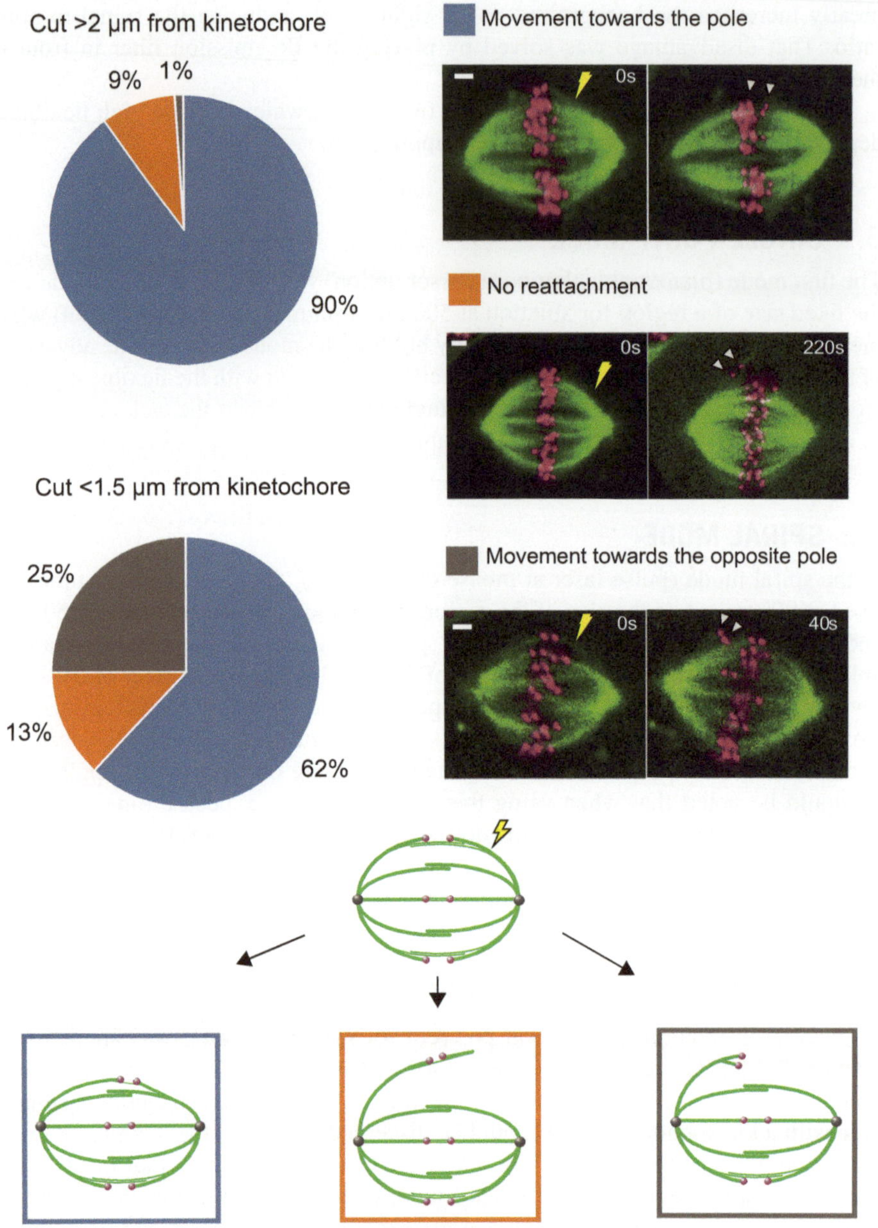

FIGURE 2 General responses to laser ablation depending on the position of the cut.

Pie charts representing the response of the ablated k-fiber to the cut performed >2 μm from the kinetochore in the first 20 s and the response of ablated k-fiber to the cut performed <1.5 μm from the kinetochore in the first 20 s (top left). Legend representing outcomes quantified in pie charts with accompanying time-lapse images of the spindle in U2OS cell

4. IMAGE ANALYSIS

4.1 CHARACTERIZATION OF GENERAL RESPONSES TO ABLATION

The main goal of the laser ablation experiment was to release one single k-fiber from the rest of the spindle. Successful k-fiber severing could be identified by observing the movement of sister kinetochores and the attached k-fiber stub (microtubule fragment attached to the kinetochore after ablation) away from the spindle long axis immediately after the severing (Fig. 2) (Kajtez et al., 2016), and later by their movement towards the pole (Elting et al., 2014; Kajtez et al., 2016; Sikirzhytski et al., 2014). After the successful laser ablation was performed >2 μm from the kinetochore, a detachment of the newly created microtubule stub from the rest of the spindle was observed during first 20 s in most of the cases (98%, n = 350 cells). In the rest of the cases, successful ablation could not be observed or the spindle collapsed and the cell presumably went to apoptosis soon after ablation (identified by blebbing from the cell membrane). After that primary response to ablation, there were several outcomes that could be observed. The most frequent outcome (90%) was the movement of the ablated k-fiber stub towards the spindle pole from which it was detached (Elting et al., 2014; Kajtez et al., 2016; Milas & Tolić, 2016; Sikirzhytski et al., 2014). However, in some cases (9%) no reattachment was observed for a longer period of time (on a minute scale). In a very few cases (1%), the movement of the ablated k-fiber stub towards the opposite spindle pole was observed (Dick & Gerlich, 2013; Sikirzhytski et al., 2014) (Fig. 2, top panel). This process resulted in a conversion of amphitelic chromosome attachment into a syntelic one. It is important to note that different outcomes previously mentioned are dependent on the position of the cut with respect to the kinetochore—the movement of the ablated k-fiber stub towards the opposite spindle pole was more frequent (25%, n = 68) when the cut was performed closer to the kinetochore (<1 μm) (Fig. 2, top panel). Most of the cells (99%, n = 418), irrespective of the position of the cut, continued with mitosis after the ablation was performed.

4.2 CHARACTERIZATION OF THE PRIMARY RESPONSE TO ABLATION IN THE FIRST 20 SECONDS

4.2.1 The amplitude of the outward movement of the sister k-fibers

The primary response to ablation is the movement of sister kinetochores and the attached k-fiber stub away from the spindle axis (Kajtez et al., 2016) (Fig. 3A). The amplitude of that movement was quantified by tracking the contours of the detached k-fiber in every frame by using the Multi-point tool in Fiji (Fig. 3B, top

expressing cenpA-GFP (magenta) and tubulin-mCherry (green) (top right). The yellow lightening sign represents the position of the cut and the white arrowheads represent the position of the pair of sister kinetochores. Scheme representing outcomes quantified in pie charts and presented in legend (bottom). Scale bar, 1 μm.

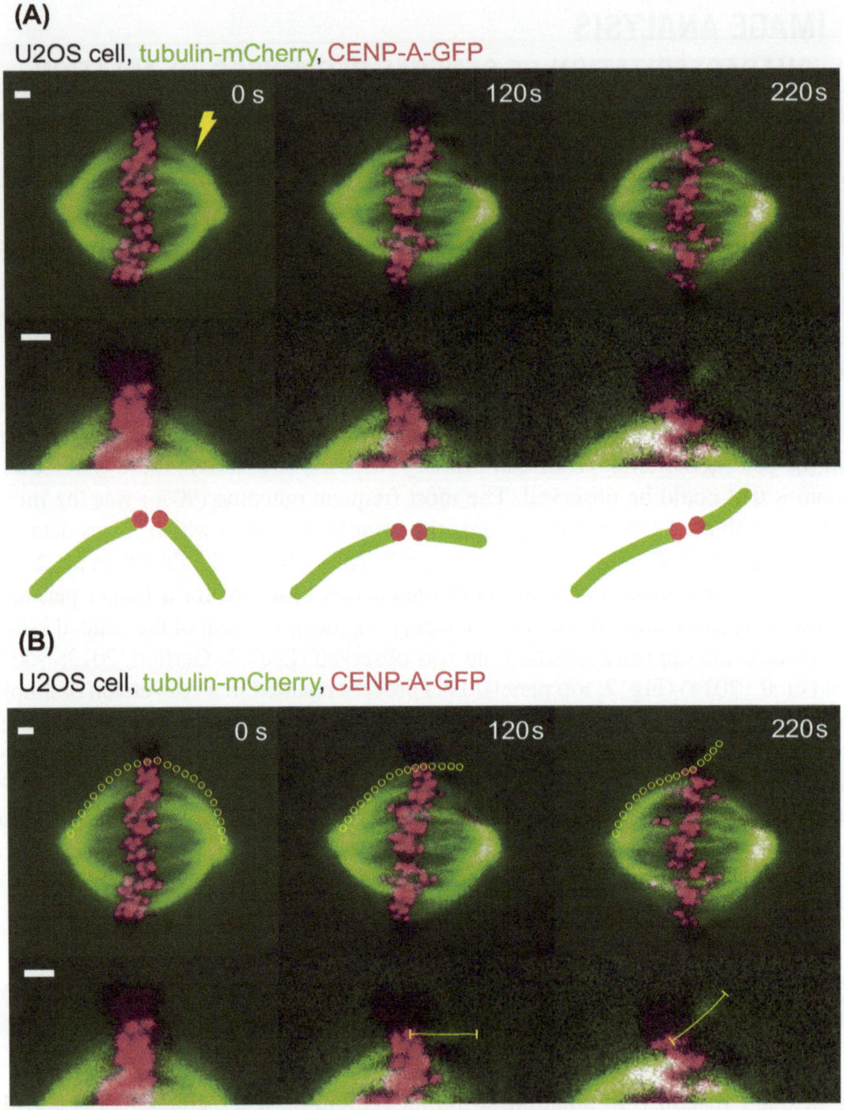

FIGURE 3 Laser ablation of microtubule bundles in U2OS cells.

(A) Time-lapse images of the spindle (top) in U2OS cell expressing cenpA-GFP (magenta) and tubulin-mCherry (green). Enlargement of the region where the ablation was performed (middle: GFP and mCherry, bottom: schemes). After the cut (yellow lighting sign), the k-fiber moved away from the long spindle axis. (B) Time-lapse images of the spindle in U2OS cell (as in A) showing tracking of contours of detached sister k-fibers in every frame by using the Multi-point tool in Fiji (yellow dots) (top) and the length of the stub by using the Segmented line tool in Fiji (yellow line) (bottom). Scale bar, 1 μm.

panel). Such measurement was done on individual imaging planes or on the maximum-intensity projection of up to three planes. A typical distance between the neighboring points on the contours was ~250 nm. The measurement was done from spindle pole along non-ablated k-fiber to the tip of the k-fiber stub (Fig. 3B, top panel). From the obtained x and y coordinates, the shape of the microtubule structure was quantitatively described. Such data also provide quantitative information on the k-fiber shape dynamics after ablation and their relation to other measured spindle parameters (for examples see Kajtez et al., 2016).

4.2.2 Characterization of the newly created microtubule minus end

The measurement of the newly created k-fiber stub was done on individual imaging planes or on the maximum-intensity projection of up to five planes. It should be noted that the microtubule fragment attached to the kinetochore remained stable during metaphase, as previously reported (Decker & Brugues, 2015) (Fig. 3B, bottom panel). Stub length was measured in every frame after the cut from the center of the corresponding kinetochore to the end of the ablated stub by using the Segmented line tool in Fiji (Fig. 3B, bottom panel). It should be noted that the measured stub length could be underestimated due to the possible partial photobleaching at the stub end during laser ablation. Also, it is important to note that because the newly created stub was not attached to any stable structure on its minus end, it was highly mobile in all directions, especially in z, which means one should be especially careful in finding all planes to which the stub moved in time.

4.2.3 Bridging fiber thickness before and after ablation

The exact number of microtubules in the bridging and the k-fiber could not be determined due to the light microscope limitations (resolution ~250 nm). Because of that, the bridging fiber thickness was measured indirectly using an approach in which the intensity ratio between the k-fiber and the bridging fiber was measured. If the number of microtubules in the k-fiber is known, the number of microtubules in the bridging fiber can be easily estimated from their ratio. Bridging fiber thickness measurement was performed on the side of the spindle where the ablation was performed, in the channel with fluorescently labeled microtubules (Fig. 4). Such measurement was done on individual imaging planes or on the maximum-intensity projection of up to three planes in the frame before and the first frame after the cut. In these planes, we could discern sister k-fibers without interference from neighboring fibers. The signal intensity of a cross-section of a bridging fiber (I_B) was measured in Fiji by drawing a 3-pixel-thick line between outermost sister kinetochores and perpendicular to the line joining the centers of the two kinetochores. The intensity profile was taken along this line and the mean value of the background signal, present in the cytoplasm around sister k-fibers (measured in Fiji by drawing a 5-pixel-thick line), was subtracted from it (Fig. 4). The signal intensity of the bridging fiber was calculated as the area under the peak closest to the kinetochores using SciDavis (Free Software Foundation, Inc., Boston, MA, USA). The width of this peak was typically 0.7 μm. The signal intensity of the k-fiber and the bridging

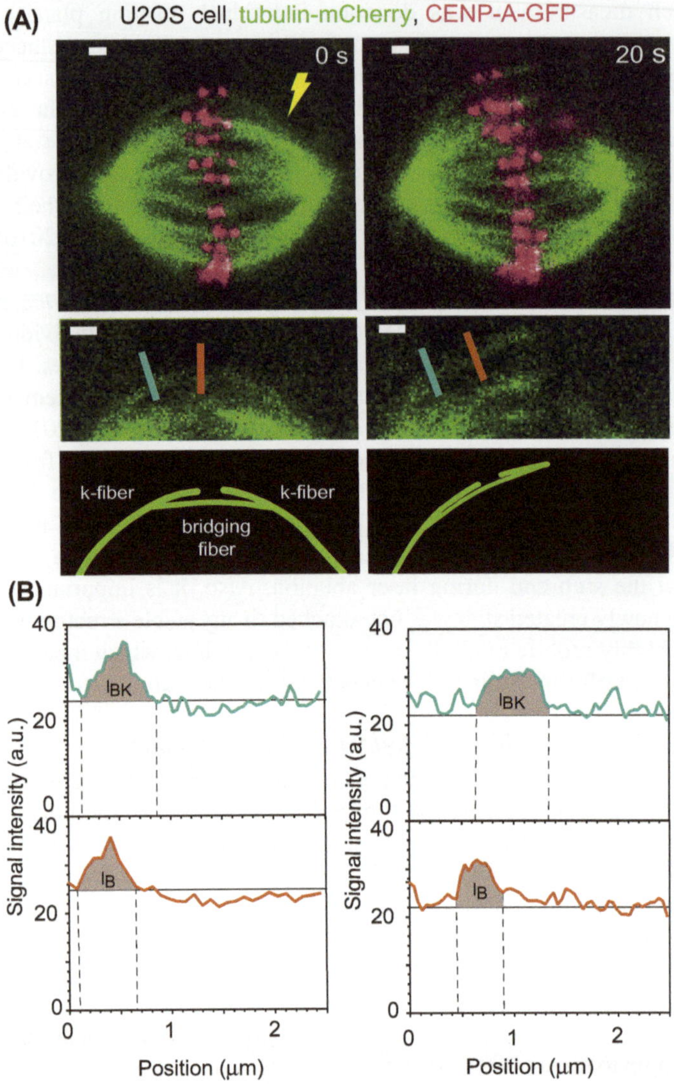

FIGURE 4 Measuring the bridging fiber thickness before and after laser ablation.

(A) Time-lapse images of the spindle (top) in U2OS cell expressing cenpA-GFP (magenta) and tubulin-mCherry (green). Enlargement of the region where the ablation was performed (top middle: GFP and mCherry, bottom middle: GFP, bottom: schemes). After the cut (yellow lighting sign), the k-fiber moved away from the long spindle axis and the bridging fiber remained in close proximity to the sister k-fibers. Orange line represents the measurement of the tubulin-mCherry signal intensity of the bridging fiber and the blue line represents the measurement of the tubulin-mCherry signal intensity of the bundle consisting of the bridging and the k-fiber. (B) Tubulin-mCherry signal intensity of the bundle fiber together (I_{BK}) was measured in a similar manner, 1 μm away from a kinetochore and perpendicular to and crossing the corresponding k-fiber. The signal intensity I_B was interpreted as the signal of the bridging fiber, and I_{BK} as the sum of the k-fiber signal and the bridging fiber signal ($I_B + I_K$) because of their lateral connection in that region (Kajtez et al., 2016). Also, due to the limited optical resolution in light microscopy which is 250 nm, it was not possible to distinguish separate bridging and k-fiber intensities within the I_{BK}.

To calculate the mean ratio of the bridging fiber to the k-fiber thickness use the following formula:

$$\frac{I_B}{I_K} = \frac{\left(\frac{I_B}{(I_{BK})}\right)}{\left(1 - \frac{I_B}{I_{BK}}\right)}$$

4.2.4 Measuring the kinetochore dynamics after ablation

Kinetochores were tracked in time using Low Light Tracking Tool (LLTT), an ImageJ plugin (Krull et al., 2014). Tracking of kinetochores in the x, y plane was performed on individual imaging planes or on maximum-intensity projections of up to three planes. The position in z direction was ignored because it had a small contribution to the kinetochore movement.

Before tracking, it is important to note that all videos designated for tracking should be saved in a separate folder. In order to obtain optimal tracking results, it was necessary to define good intensity offset in the channel with fluorescently labeled kinetochores. The intensity offset was defined by measuring the mean intensity around kinetochores in the first frame before ablation using "freehand selection" tool in Fiji. Sometimes, if photobleaching is prominent, bleach correction using Histogram Matching Method in Fiji (Miura, Rueden, Hiner, Schindelin, & Rietdorf, 2014) was done to compensate for a decrease in background intensity in time. Also, it was necessary to define EMCCD gain and electrons per A/D count of the used EMCCD camera to correct the measured flux of the object and background noise. The EMCCD-Gaussian−ML tracking algorithm method was used (Krull et al., 2014) because it yielded more precise results compared with Gaussian-ML method, especially in situations when fast movement of the tracked object occurred (on a scale of micron per frame or more). It is important to note that all tracked objects should be double checked by eye to ensure that tracking was accurate, because it was inaccurate in situations of an uneven intensity of tracked objects and in situations when multiple similar objects appeared in close

consisting of the bridging fiber and the k-fiber, I_{BK}, before ablation (left panel top) and after the ablation (right panel top). Tubulin-mCherry signal intensity of the bundle consisting of the bridging, I_B, before ablation (left panel bottom) and after the ablation (right panel bottom). Scale bar, 1 μm.

proximity. If those cases are predominant, tracking should be performed manually. σ value (standard deviation of the Gaussian used to approximate the Point Spread Function (PSF) of the tracked objects) was set to 1 to encompass just the tracked kinetochore. Also, it is important to note that all x, y, and z coordinates of the objects tracked with LLTT are given in pixels.

CONCLUSION

The presented method provides a simple, fast, and precise way of performing optical micromanipulation in vivo using NIR femtosecond laser ablation combined with SFC imaging. It provides the ability to detach a single element, such as k-fiber, from the rest of the spindle with minimal collateral damage to the rest of the spindle. Such ability is especially important in live cell imaging where any collateral damage can influence the studied processes. We have shown that after the successful ablation, cells were able to continue with normal cell division, suggesting that the method is noninvasive but efficient enough to separate only one element of the spindle. Separation of one element was used to explore structural and functional roles of individual parts of the spindle, such as the bridging fiber, which was done by quantitatively measuring post-ablation changes in shape, bridging fiber thickness, stub length and stability, and interkinetochore distance (Kajtez et al., 2016; Milas & Tolić, 2016). In addition, this method can be easily combined with other quantitative methods of microtubule dynamics such as speckle and single-molecule microscopy, photoactivation, and fluorescence recovery after photobleaching (Decker & Brugues, 2015), among others, opening the possibility to describe the relations between the different spindle substructures and the forces acting on them.

ACKNOWLEDGMENTS

The authors thank Marin Barišić and Helder Maiato for U2OS cell line; Ana Milas for imaging settings regarding spindle dynamics; Marc Koch and Alberto Zorloni for help with the setup assembly and general understanding of the whole microscopy system in the Tolić lab; Sebastian Bundschuh for his professional advices and guidelines in optimizing the system; the whole Tolić laboratory team and Nenad Pavin for helpful discussions; and Ivana Šarić for help with the drawings. This work was supported by the European Research Council (ERC).

REFERENCES

Balchand, S. K., Mann, B. J., & Wadsworth, P. (2016). Using fluorescence microscopy to study mitosis. *Methods in Molecular Biology, 1413*, 3–14. http://dx.doi.org/10.1007/978-1-4939-3542-0_1.

Botvinick, E. L., Venugopalan, V., Shah, J. V., Liaw, L. H., & Berns, M. W. (2004). Controlled ablation of microtubules using a picosecond laser. *Biophysical Journal, 87*(6), 4203–4212. http://dx.doi.org/10.1529/biophysj.104.049528.

Bringmann, H., & Hyman, A. A. (2005). A cytokinesis furrow is positioned by two consecutive signals. *Nature, 436*(7051), 731–734. http://dx.doi.org/10.1038/nature03823.

Brito, D. A., Strauss, J., Magidson, V., Tikhonenko, I., Khodjakov, A., & Koonce, M. P. (2005). Pushing forces drive the comet-like motility of microtubule arrays in Dictyostelium. *Molecular Biology of the Cell, 16*(7), 3334–3340. http://dx.doi.org/10.1091/mbc.E05-01-0057.

Cojoc, G., Roscioli, E., Zhang, L., Garcia-Ulloa, A., Shah, J. V., Berns, M. W., … Gregan, J. (2016). Laser microsurgery reveals conserved viscoelastic behavior of the kinetochore. *The Journal of Cell Biology, 212*(7), 767–776. http://dx.doi.org/10.1083/jcb.201506011.

Colombelli, J., Reynaud, E. G., Rietdorf, J., Pepperkok, R., & Stelzer, E. H. (2005). In vivo selective cytoskeleton dynamics quantification in interphase cells induced by pulsed ultraviolet laser nanosurgery. *Traffic, 6*(12), 1093–1102. http://dx.doi.org/10.1111/j.1600-0854.2005.00334.x.

Colombelli, J., Reynaud, E. G., & Stelzer, E. H. (2005). Subcellular nanosurgery with a pulsed subnanosecond UV-A laser. *Medical Laser Application, 20*(3), 217–222. http://dx.doi.org/10.1016/j.mla.2005.07.003.

Decker, F., & Brugues, J. (2015). Dissecting microtubule structures by laser ablation. *Methods in Cell Biology, 125*, 61–75. http://dx.doi.org/10.1016/bs.mcb.2014.11.004.

Dick, A. E., & Gerlich, D. W. (2013). Kinetic framework of spindle assembly checkpoint signalling. *Nature Cell Biology, 15*(11), 1370–1377. http://dx.doi.org/10.1038/ncb2842.

Dumont, S., & Mitchison, T. J. (2009). Force and length in the mitotic spindle. *Current Biology: CB, 19*(17), R749–R761. http://dx.doi.org/10.1016/j.cub.2009.07.028.

Elting, M. W., Hueschen, C. L., Udy, D. B., & Dumont, S. (2014). Force on spindle microtubule minus ends moves chromosomes. *The Journal of Cell Biology, 206*(2), 245–256. http://dx.doi.org/10.1083/jcb.201401091.

Goudarzi, M., Banisch, T. U., Mobin, M. B., Maghelli, N., Tarbashevich, K., Strate, I., … Raz, E. (2012). Identification and regulation of a molecular module for bleb-based cell motility. *Developmental Cell, 23*(1), 210–218. http://dx.doi.org/10.1016/j.devcel.2012.05.007.

Guarino, E., Cojoc, G., Garcia-Ulloa, A., Tolic, I. M., & Kearsey, S. E. (2014). Real-time imaging of DNA damage in yeast cells using ultra-short near-infrared pulsed laser irradiation. *Plos One, 9*(11), e113325. http://dx.doi.org/10.1371/journal.pone.0113325.

Kajtez, J., Solomatina, A., Novak, M., Polak, B., Vukusic, K., Rudiger, J., … Tolic, I. M. (2016). Overlap microtubules link sister k-fibres and balance the forces on bi-oriented kinetochores. *Nature Communications, 7*, 10298. http://dx.doi.org/10.1038/ncomms10298.

Khodjakov, A., La Terra, S., & Chang, F. (2004). Laser microsurgery in fission yeast; role of the mitotic spindle midzone in anaphase B. *Current Biology: CB, 14*(15), 1330–1340. http://dx.doi.org/10.1016/j.cub.2004.07.028.

Klingner, C., Cherian, A. V., Fels, J., Diesinger, P. M., Aufschnaiter, R., Maghelli, N., … Wedlich-Soldner, R. (2014). Isotropic actomyosin dynamics promote organization of the apical cell cortex in epithelial cells. *The Journal of Cell Biology, 207*(1), 107–121. http://dx.doi.org/10.1083/jcb.201402037.

Krull, A., Steinborn, A., Ananthanarayanan, V., Ramunno-Johnson, D., Petersohn, U., & Tolic-Norrelykke, I. M. (2014). A divide and conquer strategy for the maximum likelihood localization of low intensity objects. *Optics Express, 22*(1), 210–228. http://dx.doi.org/10.1364/OE.22.000210.

La Terra, S., English, C. N., Hergert, P., McEwen, B. F., Sluder, G., & Khodjakov, A. (2005). The de novo centriole assembly pathway in HeLa cells: cell cycle progression and centriole assembly/maturation. *The Journal of Cell Biology, 168*(5), 713–722. http://dx.doi.org/10.1083/jcb.200411126.

Maghelli, N., & Tolic-Norrelykke, I. M. (2010). Optical trapping and laser ablation of microtubules in fission yeast. *Methods in Cell Biology, 97*, 173–183. http://dx.doi.org/10.1016/S0091-679X(10)97010-6.

Maghelli, N., & Tolic-Norrelykke, I. M. (2011). Laser ablation of the microtubule cytoskeleton: setting up and working with an ablation system. *Methods in Molecular Biology, 777*, 261–271. http://dx.doi.org/10.1007/978-1-61779-252-6_19.

Magidson, V., Loncarek, J., Hergert, P., Rieder, C. L., & Khodjakov, A. (2007). Laser microsurgery in the GFP era: a cell biologist's perspective. *Methods in Cell Biology, 82*, 239–266. http://dx.doi.org/10.1016/S0091-679X(06)82007-8.

Maiato, H., Khodjakov, A., & Rieder, C. L. (2005). Drosophila CLASP is required for the incorporation of microtubule subunits into fluxing kinetochore fibres. *Nature Cell Biology, 7*(1), 42–47. http://dx.doi.org/10.1038/ncb1207.

Maiato, H., Rieder, C. L., & Khodjakov, A. (2004). Kinetochore-driven formation of kinetochore fibers contributes to spindle assembly during animal mitosis. *The Journal of Cell Biology, 167*(5), 831–840. http://dx.doi.org/10.1083/jcb.200407090.

Milas, A., & Tolic, I. M. (2016). Relaxation of interkinetochore tension after severing of a k-fiber depends on the length of the k-fiber stub. *Matters (Zür.)*. http://dx.doi.org/10.19185/matters.201603000025. Published online March 23.

Miura, K., Rueden, C., Hiner, M., Schindelin, J., & Rietdorf, J. (2014). *ImageJ Plugin CorrectBleach V2.0.2 [Data set], Zenodo*. http://dx.doi.org/10.5281/zenodo.30769.

Pavin, N., & Tolić, I. M. (2016). Self-organization and forces in the mitotic spindle. *Annual Review of Biophysics, 45*, 279–298. http://dx.doi.org/10.1146/annurev-biophys-062215-010934.

Raabe, I., Vogel, S. K., Peychl, J., & Tolic-Norrelykke, I. M. (2009). Intracellular nanosurgery and cell enucleation using a picosecond laser. *Journal of Microscopy, 234*(1), 1–8. http://dx.doi.org/10.1111/j.1365-2818.2009.03142.x.

Reinhardt, D., Frenz, M., Mandel, T., & Kuhlemeier, C. (2005). Microsurgical and laser ablation analysis of leaf positioning and dorsoventral patterning in tomato. *Development, 132*(1), 15–26. http://dx.doi.org/10.1242/dev.01544.

Rumpf, C., Cipak, L., Schleiffer, A., Pidoux, A., Mechtler, K., Tolic-Norrelykke, I. M., & Gregan, J. (2010). Laser microsurgery provides evidence for merotelic kinetochore attachments in fission yeast cells lacking Pcs1 or Clr4. *Cell Cycle, 9*(19), 3997–4004. http://dx.doi.org/10.4161/cc.9.19.13233.

Shen, N., Datta, D., Schaffer, C. B., LeDuc, P., Ingber, D. E., & Mazur, E. (2005). Ablation of cytoskeletal filaments and mitochondria in live cells using a femtosecond laser nanoscissor. *Mechanics & Chemistry of Biosystems, 2*(1), 17–25. http://dx.doi.org/10.3970/mcb.2005.002.017.

Sikirzhytski, V., Magidson, V., Steinman, J. B., He, J., Le Berre, M., Tikhonenko, I., … Khodjakov, A. (2014). Direct kinetochore-spindle pole connections are not required for chromosome segregation. *The Journal of Cell Biology, 206*(2), 231–243. http://dx.doi.org/10.1083/jcb.201401090.

Simunić, J., & Tolić, I. M. (2016). Mitotic spindle assembly: building the bridge between sister k-fibers. *Trends in Biochemical Sciences, 41*(10), 824–833. http://dx.doi.org/10.1016/j.tibs.2016.07.004.

Steinmeyer, J. D., Gilleland, C. L., Pardo-Martin, C., Angel, M., Rohde, C. B., Scott, M. A., & Yanik, M. F. (2010). Construction of a femtosecond laser microsurgery system. *Nature Protocols Other Titles: Protocols, 5*(3), 395–407. http://dx.doi.org/10.1038/nprot.2010.4.

Stiess, M., Maghelli, N., Kapitein, L. C., Gomis-Ruth, S., Wilsch-Brauninger, M., Hoogenraad, C. C., … Bradke, F. (2010). Axon extension occurs independently of centrosomal microtubule nucleation. *Science, 327*(5966), 704–707. http://dx.doi.org/10.1126/science.1182179.

Tolić, I. M., & Pavin, N. (2016). Bridging the gap between sister kinetochores. *Cell Cycle, 15*(9), 1169–1170. http://dx.doi.org/10.1080/15384101.2016.1157976.

Tolic-Norrelykke, I. M., Sacconi, L., Thon, G., & Pavone, F. S. (2004). Positioning and elongation of the fission yeast spindle by microtubule-based pushing. *Current Biology: CB, 14*(13), 1181–1186. http://dx.doi.org/10.1016/j.cub.2004.06.029.

Vogel, A., Lorenz, K., Horneffer, V., Huttmann, G., von Smolinski, D., & Gebert, A. (2007). Mechanisms of laser-induced dissection and transport of histologic specimens. *Biophysical Journal, 93*(12), 4481–4500. http://dx.doi.org/10.1529/biophysj.106.102277.

Vogel, A., Noack, J., Hüttman, G., & Paltauf, G. (2005). Mechanisms of femtosecond laser nanosurgery of cells and tissues. *Applied Physics B, 81*(8), 1015–1047. http://dx.doi.org/10.1007/s00340-005-2036-6.

Vogel, S. K., Pavin, N., Maghelli, N., Julicher, F., & Tolic-Norrelykke, I. M. (2009). Self-organization of dynein motors generates meiotic nuclear oscillations. *PLoS Biol, 7*(4), e1000087. http://dx.doi.org/10.1371/journal.pbio.1000087.

Watanabe, W., Arakawa, N., Matsunaga, S., Higashi, T., Fukui, K., Isobe, K., & Itoh, K. (2004). Femtosecond laser disruption of subcellular organelles in a living cell. *Optics Express, 12*(18), 4203–4213. http://dx.doi.org/10.1364/opex.12.004203.

Yanik, M. F., Cinar, H., Cinar, H. N., Chisholm, A. D., Jin, Y., & Ben-Yakar, A. (2004). Neurosurgery: functional regeneration after laser axotomy. *Nature, 432*(7019), 822. http://dx.doi.org/10.1038/432822a.

CHAPTER

Fluorescence eXclusion Measurement of volume in live cells

6

C. Cadart*, E. Zlotek-Zlotkiewicz*, L. Venkova*, O. Thouvenin[§], V. Racine[¶], M. Le Berre*, S. Monnier*[,1], M. Piel*[,1]

**Institut Curie/Institut Pierre-Gilles de Gennes, Paris, France*

[§]Institut Langevin/ESPCI Paris Tech, Paris, France

[¶]QuantaCell, Pessac, France

[1]Correspoding authors: E-mail: sylvain.monnier@univ-lyon1.fr; matthieu.piel@curie.fr

CHAPTER OUTLINE

Abstract

Volume is a basic physical property of cells; however, it has been poorly investigated in cell biology so far, mostly because it is difficult to measure it precisely. Recently, large efforts were made to experimentally measure mammalian cell size and used mass, density, or volume as proxies for cell size. Here, we describe a method enabling cell volume measurements for single living cells. The method is based on the principle of fluorescent dye exclusion and can be easily implemented in cell biology laboratories. As

Methods in Cell Biology, Volume 139, ISSN 0091-679X, http://dx.doi.org/10.1016/bs.mcb.2016.11.009

this method is very versatile, it can be used for cells of different sizes, adherent or growing in suspension, over several cell cycles and is independent of cell shape changes. The method is also compatible with traditional cell biology tools such as epifluorescence imaging or drug treatments.

INTRODUCTION

The role of cell volume regulation in various cellular processes has recently emerged as a topic of interest. In confined environments, it was shown that tumor cell migration depends on osmotic control and volume regulation through ion pumps' activity and localization (Stroka et al., 2014). Recent findings also demonstrated that during cell division, mammalian cells tightly regulate their volume and exhibit a transient swelling, through osmotic processes controlled by ion pumps (Son et al., 2015; Zlotek-Zlotkiewicz, Monnier, Cappello, Le Berre, & Piel, 2015). Local or global cell volume changes could thus be a way of generating forces in cells. Therefore, osmotic regulation could be used by mammalian cells, just like in plants or yeasts, to expand or change shape (Cosgrove, 2005). The regulation of volume has been studied from a theoretical point of view (Jiang & Sun, 2013; Mitchison, Charras, & Mahadevan, 2008; Tao & Sun, 2015) but experimental test of these theoretical works is mostly lacking. At longer timescales, mammalian cell growth and cell-size homeostasis throughout the cell division cycle are currently hotly debated (Ginzberg, Kafri, & Kirschner, 2015). Part of the experimental studies suggested that cell volume is a biologically meaningful measure of cell size (Conlon & Raff, 2003; Godin et al., 2010; Son et al., 2012; Tzur, Kafri, LeBleu, & Lahav, 2009), while others favor mass (Mir et al., 2011; Park et al., 2010; Sung et al., 2013). The investigation of these questions is currently limited by the difficulty in measuring single-cell volume accurately over long periods of time.

Indeed, volume measurement on cells that constantly modify their shape is a very challenging task. For decades, the only commercially available technique was the Coulter counter. This technique yields high-throughput measurements based on impedance. It is however limited to cells in suspension and it is not possible to repeat the measurement on the same single cell (Bryan, Engler, Gulati, & Manalis, 2012; Gregg & Steidley, 1965). More recently, cell shape reconstruction from z-stacks of confocal images of labeled membrane was proposed (Boucrot & Kirchhausen, 2008). Yet, this technique requires repeated exposures to excitation light and is therefore not suitable for experiments over long time periods. Moreover, accuracy of volume measurements is limited by labeling artifacts and cell-shape reconstruction methods (Boucrot & Kirchhausen, 2008; Ernest, Habela, & Sontheimer, 2008; Habela & Sontheimer, 2007). Finally, the suspended microchannel resonator (SMR) allows very high precision measurement of cell mass, density, and volume but is limited to cells in suspension and requires very specific instrumentation (Bryan et al., 2013; Son et al., 2015).

Here, we describe a method to measure the volume of single cells using simple microfabricated cell culture chambers. This versatile technique is based on the principle of fluorescence exclusion (Bottier et al., 2011; Gray, Hoffman, & Hansen, 1983; Zlotek-Zlotkiewicz et al., 2015) and allows the measurement of volume of suspended or adherent cells with sizes ranging from few microns to several tens of microns. It allows studying phenomena ranging from seconds (drug treatments, osmotic shocks), minutes or hours (mitosis) up to several days (cell division cycles).

1. METHOD

1.1 OVERVIEW

Cells are cultured in a poly(dimethylsiloxan) (PDMS) chamber of controlled height with regular culture media supplemented with a fluorescent dye coupled to dextran as shown in Fig. 1A. Dextran is a polysaccharide preventing the fluorescent dye from passively entering the cells. Cells thus exclude fluorescence and appear dark on a classic fluorescence image (Fig. 1A and B). The fluorescence displaced by a cell is linearly proportional to the cell volume (Bottier et al., 2011; Zlotek-Zlotkiewicz et al., 2015). It is important to note that cells with high endocytic activity are not suitable for this method as dextran would be taken up inside the cell and the fluorescence would not be excluded. However, *fluorescence exclusion measurement* (FXm) can then be used to estimate the endocytosed volume, by sequentially changing the color of the excluded extracellular dye. It is also possible to flow the fluorescent dye only for short time using microfluidics, thus avoiding the accumulation of the dye in the cells, but this method is not presented here. Overall, for most cell types, minimizing the cell stress prevents fluorophore endocytosis. Careful handling of the cells is therefore important for reliable and reproducible results.

1.2 PROTOCOL

1.2.1 General comments

This protocol describes the preparation steps for one chamber to be assembled on a glass bottom petri dish. The dish can be homemade or commercial, 6-well glass bottom plates can also be used for multiplexing.

1.2.2 Material

- Mold
- Hole puncher (diameter: 0.75 or 1 mm)
- Bottom glass petri dishes (35 mm, from WPI or Ibidi)
- PDMS and curing agent (Sylgard 184 or RTV-615)
- PBS (1×)
- Cell culture medium without phenol red
- Dextran conjugated with fluorescent dye in PBS (FITC, Alexa488, Alexa-647… from LifeTechnologies or Sigma Aldrich)

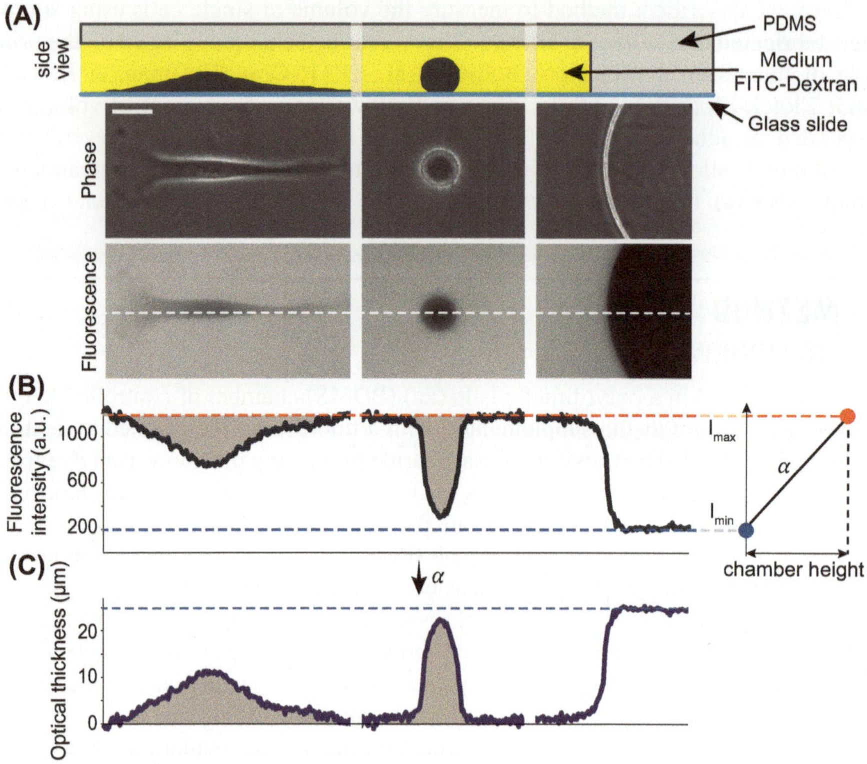

FIGURE 1 Principle of volume measurement.

(A) Cells are placed in poly(dimethylsiloxan) chambers of calibrated height set by pillars. Medium is supplemented with FITC-Dextran. Bottom picture: cells exclude fluorescence on epifluorescence images (Scale bar 20 μm). (B) The fluorescence profile corresponding to the *dotted line* in (A): maximum and minimum of fluorescence intensity correspond to chamber maximal height (background) and zero height (pillar), respectively. Right: these values are used to calibrate the signal and calculate the optical thickness of the cells. (C) Finally, cell volume is obtained by integrating the total fluorescence intensity over the cell area.

- Trypsin or EDTA
- Scalpel or razor blade
- Plasma Cleaner
- PDMS Oven between 65 and 80°C
- Cell culture hood

1.2.3 Fabrication of the PDMS chip

1.2.3.1 Rules for mold design

Molds can be fabricated by various microfabrication technics (photolithography, micromachining, ...). Three major principles should be taken into account when designing a chamber. First, the height of the chamber should be adapted according to the maximal size of the measured cells. Indeed to increase the contrast of the excluded fluorescence the ceiling should be as low as possible, but it should also be larger than the maximum cell height to prevent cells from being confined. Second, keeping a constant height throughout the chamber is of high importance as any error on the height will lead to the same error in volume calculation. Thus, the fabrication technique has to be chosen depending on the size of the cell/chamber and the precision required. The pillars keep the chamber height constant and their spacing should thus be adapted to the height of the chamber and the stiffness of the PDMS ceiling that depends on multiple parameters (i.e., thickness, brand, curing time). As a general rule, the distance between two pillars should not exceed 15 times the height of the chamber. Third, there should be at least one pillar in each field of view in order to calibrate the fluorescent signal, so the distance between the pillars should be adapted according to the field of view and the magnification that will be used.

1.2.3.2 Chamber fabrication

1. Mix PDMS with its crosslinker (1:10).
2. Pour onto the mold and degas to remove all air bubbles from the PDMS. The chip thickness should range between 2 and 4 mm.
3. Cure PDMS for 2 h in the oven between 65 and 80°C.
4. Punch 0.75 or 1-mm diameter inlets and outlets in the PDMS and cut the PDMS chip so that it fits in the glass part of the petri dishes. Narrow inlets (i.e., 0.75 mm) prevent fast diffusion of the fluorescent probe in the media outside the chip and are thus preferable for long imaging but the type of inlet can be adapted to the type of experiment.

1.2.3.3 PDMS chamber preparation: Day−1 (see Fig. 2)

1. Clean the PDMS chip with isopropanol and dry with an air gun (optional). Use a scotch tape to remove residual dusts.
2. Activate the structured side of the PDMS chip and the glass surface for 30 s in the plasma cleaner (Fig. 2A).
3. Put the PDMS chip on the glass surface immediately after plasma treatment and check carefully that the chamber and the pillars are properly bonded. A change in the reflection of light can be noticed by eye when PDMS and glass are bonded. If needed, use a soft tool (i.e., pipette tip or plastic tweezers) to gently press on the chamber to force the contact between the glass and the pillars. Additional 15 min incubation at 65−80°C reinforces the binding (optional).
4. Within less than 15 min after plasma treatment and under the hood, coat the chamber with fibronectin (50 μg/mL in PBS) by adding a drop of 10−20 μL over one of the inlets. Since the surfaces are still hydrophilic from the plasma activation, the drop will spread in the chamber easily (Fig. 2B). It is also possible to store the chips after plasma bonding, but then the chamber should be put under vacuum for 5 min and activated a second time with the plasma cleaner

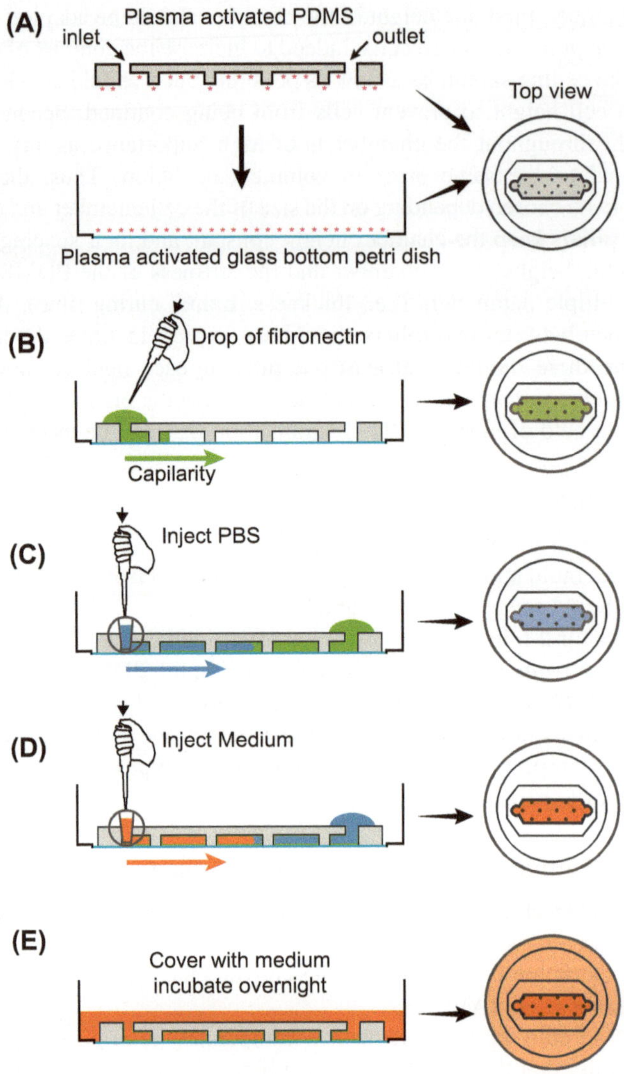

FIGURE 2 Preparation of volume chamber.

(A) Bind the poly(dimethylsiloxan) (PDMS) chamber to glass-bottom petri dish after plasma activation (30 s). (B) Apply surface treatment (fibronectin incubated for 30 min) directly after the bonding while the PDMS and glass are still hydrophilic. (C) Wash with PBS. (D) Inject medium and (E) cover the chip completely with medium before incubating overnight.

to make the surfaces hydrophilic and ensure a proper spreading of the fluid in the chamber. The plasma treatment step also sterilizes the dishes and the remaining steps can be performed under the hood in sterile conditions appropriate to cell culture.

5. Incubate for 30 min to 1 h in humid atmosphere to prevent fibronectin from drying out (i.e., add PBS or media around the PDMS chip or in the incubator).
6. Under the hood, wash with PBS and then with culture medium (Fig. 2C and D) and cover the chip with medium (Fig. 2E). Incubate overnight. Air bubbles can be generated along the preparation steps; if they are small enough (diameter < 1 mm), they should disappear during the incubation overnight. Using prewarmed media also reduces the appearance of bubbles when the chambers are put at 37°C.
   Notes:
   - This protocol can also be adapted to cells in suspension by coating the chamber with PLL-*g*-PEG (1×) rather than fibronectin to impede cell adhesion (Zlotek-Zlotkiewicz et al., 2015).
   - In order to avoid mechanically stressing the cells, one should minimize the pressure applied in the chamber during the several steps of injection.
   - If the experiment requires the use of hydrophobic drugs, an overnight incubation of the chip with medium already containing the drug should be performed to saturate the PDMS with the drug and limit drug depletion during the experiment. The concentration of hydrophobic drugs should be optimized within the chamber since an unknown amount of drugs could still be depleted from the chamber due to hydrophobicity and porosity of the PDMS material.

1.2.4 Cells' injection: Day 2 (Fig. 3)

1. Prewarm the culture medium and the Trypsin or EDTA. In case of CO_2-dependent medium, it is even better to preincubate the medium in the incubator to allow its equilibration.
2. Remove all medium around the chambers in the dish and dry the surface of the chamber with an aspirating pipette and inject medium to wash the chamber (Fig. 3A).
3. Detach cells in culture with TrypLE, Trypsin, or EDTA depending on the cell type and the time required for adhesion, using EDTA results in shorter readhesion times for most cell types.
4. Resuspend cells with warm media and centrifuge them to remove the trypsin and resuspend well to a final concentration ranging from approx. 0.5 to 2×10^6 cells/mL (Fig. 3B). Cell concentration has to be adapted to the final density required for the experiment. Obtaining single-cell volumes using this method can be done only on cells that are clearly separated. The cell density should thus be adapted to the requirement of the cell line/experiment (i.e., some cell lines require high density to properly proliferate while fast migrating cells

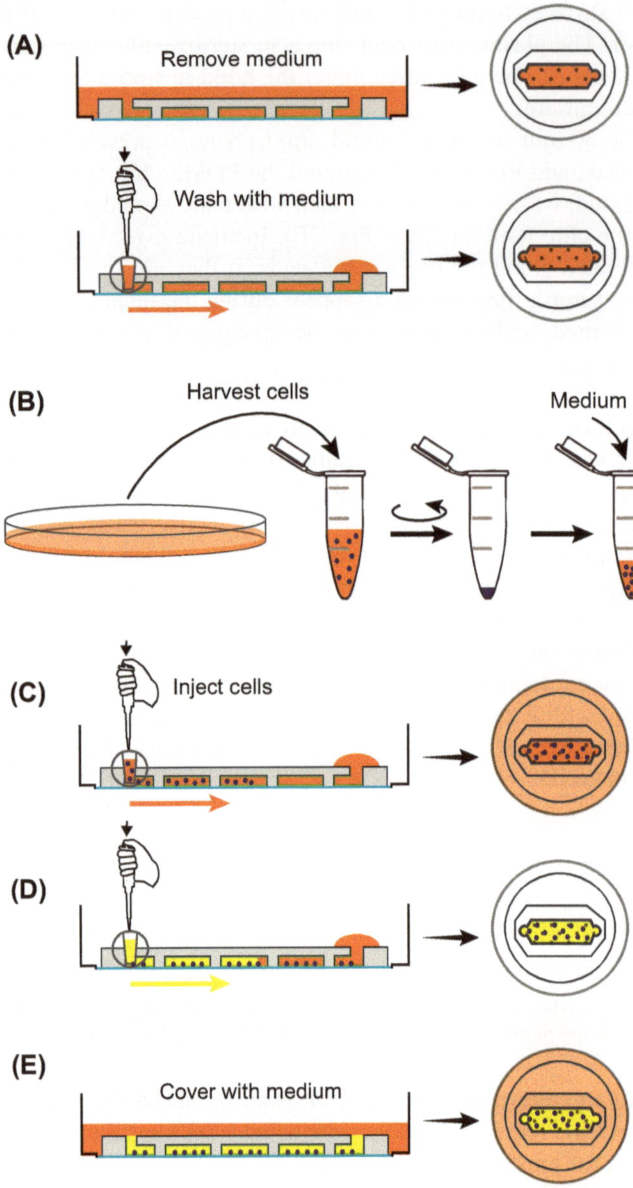

FIGURE 3 Cells' injection into the chamber.

(A) Remove medium on top and wash the chamber with fresh medium. (B) Harvest cells, spin and resuspend in fresh medium to the concentration needed. Collect a small volume of cells (typically 10 μL). (C) Inject cells, cover with medium, and incubate few hours for adhesion. (D) Change the medium to the medium supplemented with FITC-dextran. (E) Wash the medium on top of the chip and cover completely with the medium without FITC dextran.

at too high density will constantly bump into each other or migrate out of the observation field). Moreover, it is important to resuspend well the cells in order to avoid cellular aggregates that cannot be measured.

5. Inject the cells slowly to prevent mechanical stress. Check cell density under a bench microscope. The solution of cells can be concentrated or diluted according to the density observed in the chamber and reinjected.
6. Once the cell density is correct, cover completely the chamber with medium to stop the flow and incubate a few hours for the cells to adhere properly (Fig. 3C). The incubation time depends on the approach used to detach the cells and on the cell line itself but is typically ranging from 1 to 4 h.
7. For one chip; prepare 10 μL of medium supplemented with FITC-Dextran to a final concentration of 1 g/L. This solution can be prepared when cells are injected and stored in the incubator to equilibrate the medium.
8. Remove the medium above the chip and gently inject the prewarmed and equilibrated medium in the chip (Fig. 3D). One should see the dextran-colored medium getting out of the outlet.
9. Suck out all medium in the petri dish and replace by new medium completely covering the chamber (Fig. 3E). This step is very important: first, to wash dextran remaining outside of the chamber and avoid parasite fluorescence signals coming from outside the chamber; second, to completely cover the chip with the medium to prevent evaporation during the image acquisition under the microscope since PDMS is porous to gaz.

1.3 IMAGING

The imaging should be performed with a simple epifluorescence microscope, adapted for live cell imaging, with a chamber with controlled temperature, humidity, and CO_2. For long experiments, keeping the atmosphere humid is particularly important in order to prevent evaporation that could lead to changes of osmolarity in the volume measurement chamber.

The acquisition can be performed with various objectives with different magnifications and numerical apertures (NA). However, it is important to keep in mind that the FXm method is based on the collection of the light over the entire height of the chamber (see data analysis), which depends on the depth of field provided by each objective. The depth of field of an objective, which defines the range of its acceptable focus, scales with the inverse square of its numerical aperture. Nevertheless, we could demonstrate theoretically and experimentally that the volume measurement does not depend much on the depth of field or on the axial position of focalization (as shown respectively in Figs. 4 and 5A). Indeed, with an epifluorescence microscope, all fluorescence light is integrated on the camera, so even the out-of-focus information can be retrieved in the signal. The out-of-focus fluorescence is simply diffracted on several pixels of the camera around the real position of the object. If the area over which fluorescence intensity is integrated is large enough (see Section 1.4), one can therefore collect all the information coming from this

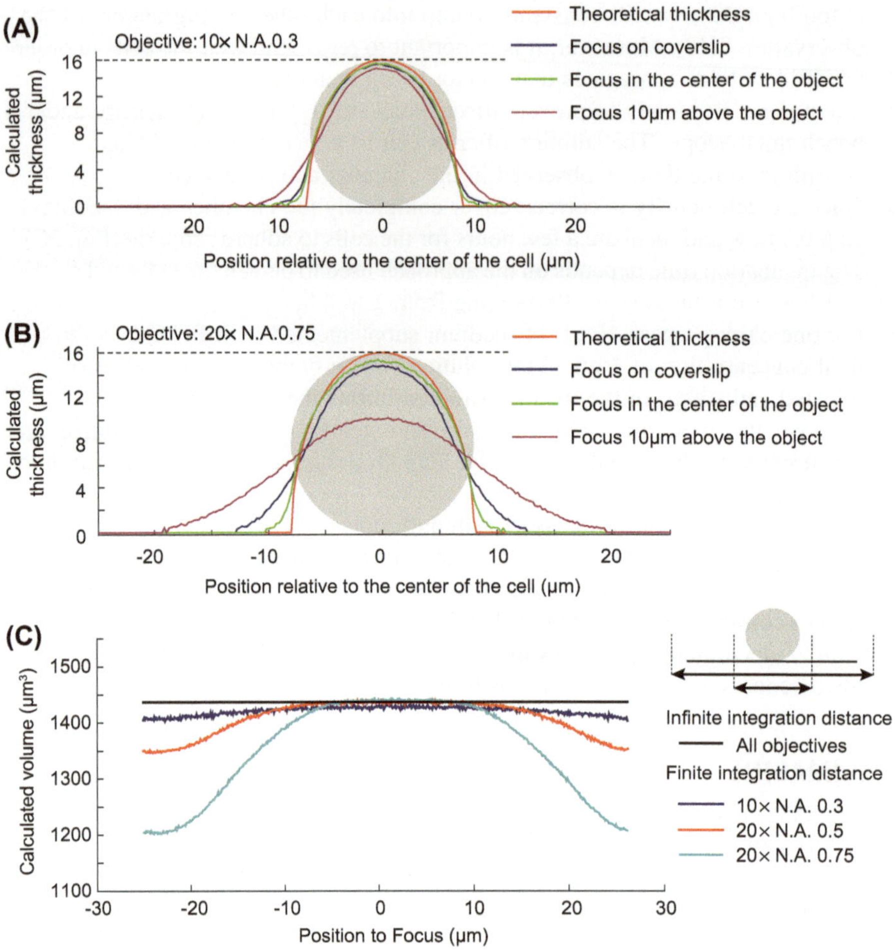

FIGURE 4 Simulated thickness and volume measured with FXm.

Results of a home-made 2D simulation showing the effect of a change of the depth of field on the calculated thickness and volume of the cell. The curves show the simulated thickness of a round theoretical cell of 16 µm in diameter in a 22-µm high chamber at multiple foci. Simulations were performed for 10× 0.3NA objective (A) and 20× 0.75NA objective (B). *Grey disk* represent the simulated object. (C) Volume calculation for a sphere of radius Rc = 8 µm from 2D simulations. The surface of the 2D cell was calculated using an infinite integration distance and with a finite distance of 20 µm. The surface was multiplied by 4/3 × Rc to obtain the volume.

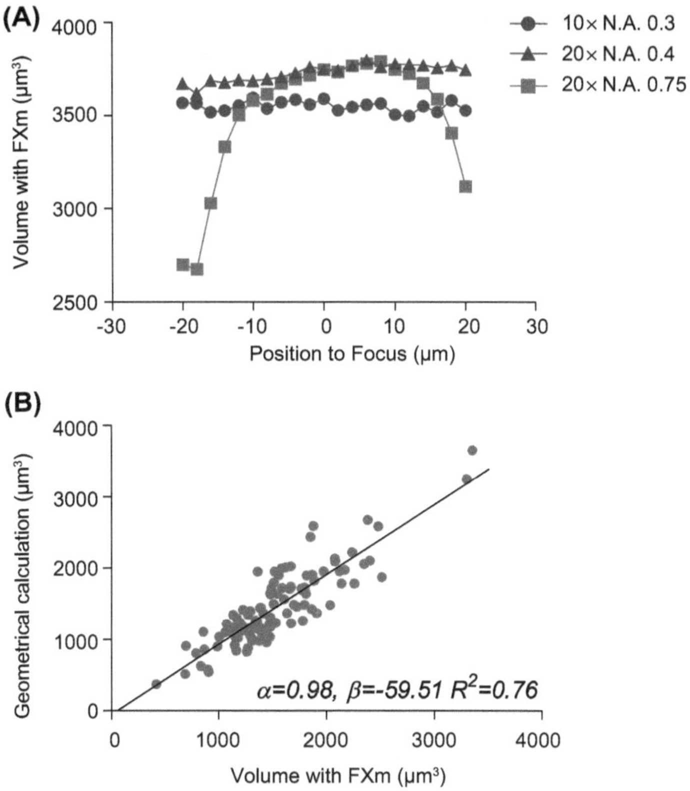

FIGURE 5 Experimental validation of the setup.

(A) Volume measured using FXm of the same cell obtained as a function of Z-position for objectives with different magnification and numerical aperture. Objectives with lowest numerical aperture yield robust volume measurement over a 20-μm range below or above the focal plan. (B) On round cells, comparison of volumes measured with FXm versus volumes calculated from the measurement of the cell diameter. The robust linear fit shows a good correlation between the two measurements. Measurements were made in a 25.2-μm high chamber with an objective LD 20× NA0.4.

cell. However, having a depth of field too small with respect to the object or the chamber height will have two main effects. First, the accuracy on a single pixel height measurement drastically reduces, since the out-of-focus fluorescence has diffracted on the neighbor pixels (Figs. 4 and 5A). Second, it is then possible to integrate information from a neighboring out-of-focus cell, dust, or pillar, which will lead to an overestimation of the calculated volume.

Fig. 4 presents the results of a home-made simulation in two dimensions of a spherical cell (diameter: 16 μm) in a volume measurement chamber (height: 22 μm). The thickness of the cell obtained by FXm was simulated for multiple

Z-positions. It shows that the calculated thickness at a given pixel can be largely misestimated when part of the cell or the chamber is out of focus. In any case, the 2D volume calculated over infinite integration distance is constant and equal to the theoretical value of 201 μm^2, which is the theoretical surface of the cross section of this round cell (Fig. 4C). The simulations show that for a finite (20-μm) integration distance, the calculated volume drops for largely out of focus images (Fig. 4C), as observed experimentally (Fig. 5A).

In conclusion, fluorescence exclusion yields accurate measurements of volumes if calculated over large enough areas containing no other object than the cell measured, while it can be wrong for single-pixel height measurements, depending on the objective's depth of field.

In order to experimentally validate a given setup (height of the chamber and objective), control measurements comparing FXm with another volume measurement method can be made. So far, four different objectives were validated using giant unilamellar vesicles (GUVs) as references: 10× NA0.3, 10× NA0.5, 20× NA0.5, and 20× NA0.75 (Zlotek-Zlotkiewicz et al., 2015) and three others have been validated with AFM measurements: 40× NA0.75 air, 40× NA 1 Oil, and 100× NA1.25 Oil (Bottier et al., 2011). Fig. 5 shows two examples of controls that can be rapidly made to estimate the accuracy and the limitations of a given setup (i.e., chamber height, objectives).

The homogeneity of the fluorescence illumination is also important for the precision of the measurement, as it will ease image segmentation and signal calibration. The light path must be clean, without any dusts obstructing the fluorescence (see Section 1.4).

1.4 DATA ANALYSIS (FIG. 6)

1.4.1 Principle of volume calculation

The FXm measurement is based on the inverse linear relationship between intensity of fluorescence and the thickness or height of the sample over which fluorescence is measured:

$$I_{max} - I_{min} = \alpha \cdot h_{chamber} \tag{1}$$

I_{max} corresponds to the intensity measured where fluorescence is present over the entire height of the chamber whereas I_{min} corresponds to the minimum intensity at the center of the pillar of the chamber. The linearity of this relationship has been verified in Zlotek-Zlotkiewicz et al.. The extraction of α enables the calculation of the optical thickness of an object excluding the fluorescence at each pixel. As discussed in the Section 1.3, this optical thickness can misestimate the real local thickness, but integrating this signal over the area S around the cell gives accurate volumes:

$$V_{cell} = \iint_S \frac{I_{max}(x,y) - I(x,y)}{\alpha} \tag{2}$$

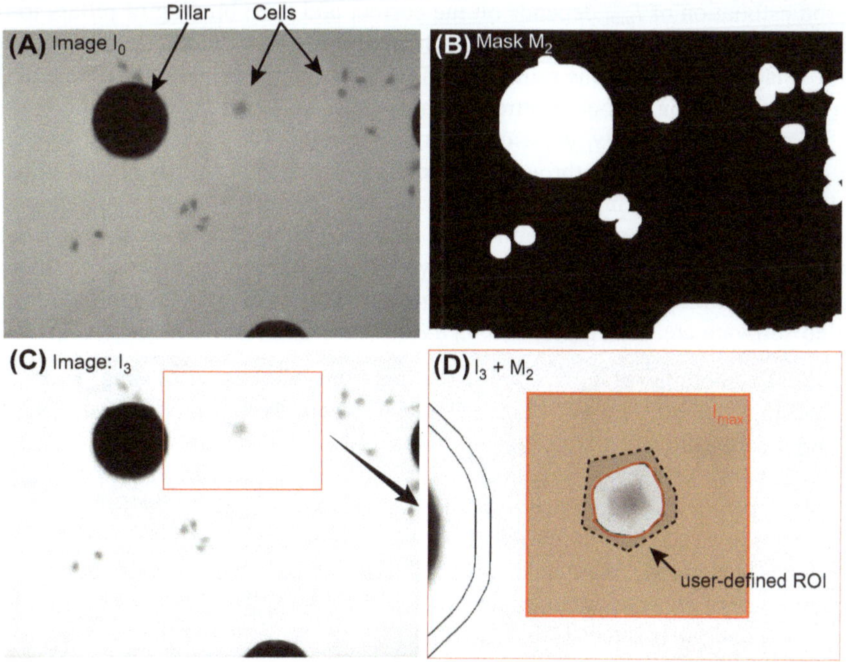

FIGURE 6 Volume calculation procedure.

(A) Raw image of a typical field observed with a 10× objective NA0.3. A pillar sustaining the ceiling is on the left and parts of two other pillars can be observed on the edges of the field. HeLa cells appear in grey. (B) The mask M_2 calculated to exclude cells, pillars, and dusts from the background. (C) Final image with renormalized background (I_3). (D) Close up from (C) showing the ROI integrated to calculate the volume of the cell, and in red, the area around the cell used for local calculation of I_{max}.

1.4.2 Description of the Matlab software from QuantaCell for image analysis

Image analysis is performed using a Matlab software first designed in the lab and optimized by QuantaCell (a company specialized in quantitative image analysis for life sciences). The software proceeds through two main steps. First, it processes raw images and normalizes them. Second, manual or automated cell tracking is performed through a user-friendly and versatile interface that is suitable to a wide range of measurements (i.e., short time on multiple cells or lineage tracking over long periods of time).

1.4.2.1 Background normalization

The accuracy of volume calculation highly depends on the accuracy of the estimation of I_{max} and I_{min} that yields α (Eq. 1). A good estimation of I_{max} first depends on the acquisition of good images with a clean background (devoid of dusts), homogeneous fluorescent probe in the chamber and homogeneous fluorescence illumination.

A good estimation of I_{min} depends on the correct and even binding of pillars to the glass bottom of the chamber. However, when all these precautions have been taken before image acquisition, one source of inhomogeneity will remain: the image of the lamp. For this reason, a good posttreatment protocol to normalize background intensity is crucial. Importantly, it seemed that in our setup, the image of the lamp can vary a bit through time or through positions in the chamber (possibly because of the PDMS structure). Thus, we chose to calculate the image of the lamp from our images for each frame rather than from a reference image made in an empty field. This alternative method might however give satisfying results in other experimental setups. Briefly, the different steps of the background normalization protocol in our Matlab software are, for each frame, I_0 as shown (Fig. 6A):

1. A first smoothing of I_0 is obtained using a morphological closing. From the resulting image I_1, a mask M_1 is calculated where dark pillars or dusts are excluded by filtering pixels below a threshold Th_1. The value of Th_1 is calculated as follows: $Th_1 = \text{percentile}(I_1, 0.1) \times 0.8 + \text{percentile}(I_1, 50) \times 0.2$. The percentile ($I_0$, 0.1) and percentile ($I_0$, 50) corresponds to the 0.1 and 50 percentiles of the fluorescence intensity distribution, respectively.
2. From this mask M_1, the mean value I_{min} of the pillar intensity is calculated over the center of all segmented pillars (to avoid border effects, an erosion step is applied and pillar intensity is estimated only at the center of the pillar on a minimum of 1000 pixels).
3. B_1 is obtained by applying M_1 to I_1 and filling in the holes through performing an iterative Gaussian blur of the background signal around each hole. B_1 thus corresponds to a first estimation of the background.
4. I_2 is then obtained as follows: $I_2 = (I_1 - I_{min})/B_1$. This image corresponds to a cleaned image of I_0 and is used to calculate a new mask.
5. From I_2, a second mask M_2 excluding the pillars and the cells is calculated, using the same filtering strategy as in Step 1 with the threshold Th_2. It is indeed only possible to reliably detect cells on a normalized I_2 image. The threshold Th_2 is calculated as follows: $Th_2 = \text{percentile}(I_2, 50) \times 0.5 + \text{percentile}(I_2, 1) \times 0.5$. Then M_2 is a merge of M_1 and M_2: $M_2 = \max(M_1, M_2)$ (Fig. 6B).
6. B2 is obtained by applying M_1 to I_0 with the same iterative Gaussian blur used as in Step 2 to fill in the holes.
7. The final renormalized image I_3 is then obtained: $I_3 = I_0/B_2$ (Fig. 6C).

1.4.2.2 Cell tracking and volume measurement

For each cell, the calibration of fluorescence intensity as a function of thickness is performed as in Eq. (1) with a value of I_{max} estimated locally (Fig. 6D). It is however true that after the normalization procedure that produces a very flat background, this additional precaution might be considered optional.

To calculate cell volume, the tracking software performs the following steps:

1. The user or the automated tracking protocol defines a region of interest (ROI) around each cell to be analyzed.

2. I_{max} is calculated locally around each cell: a fixed-sized window S centered on the ROI is defined see Fig. 6D (this area has been set to 200 × 200 pixels around the cell for a 10× binning one image). M_2, the mask excluding cells and pillars, is applied to S. The median value of the remaining background fluorescence intensity is then calculated. This step ensures that median is estimated on a large enough number of background pixels.
3. I_{min} has been previously calculated and stored as explained in Step 2 of the background normalization procedure.
4. Using these calibration parameters, cell volume is then calculated by integrating fluorescence intensity on the ROI defined (see Eq. 2 and Fig. 6D).

2. FXm PITFALLS AND LIMITATIONS

The FXm method is excellent for relative volume measurements such as those performed during mitosis (Zlotek-Zlotkiewicz et al., 2015). However, in order to ensure good absolute measurements, one must pay attention to five categories of parameters.

- The accuracy of the calibration of the optical thickness of a pixel will depend on both the flatness of the roof of the chamber and the even binding of the pillars to the bottom of the chamber. These two parameters must be even throughout the device and the time of the experiment.
- Absolute volume measurement requires that fluorescence intensity signal is homogeneous in the chamber, which can be ensured by a good diffusion of the fluorescent probe in the chamber and a homogeneous fluorescence illumination. To correct for the inevitable remaining inhomogeneity of illumination, a good image analysis protocol is important.
- The axial/height resolution will depend on the signal-to-noise ratio, which depends here on the maximal fluorescence intensity. The three adjustable parameters for this are: the illumination parameters, the concentration of the fluorescent probe that will yield a homogeneous background signal, and the difference between the height of the chamber and the height of the object measured.
- As explained in Section 1.3, the single-pixel thickness is often misestimated, and the volume can be as well when the depth of field is too small compared to the object's size. The acceptable height over which fluorescence intensity extraction is reliable depends on the objective used.
- In addition to these parameters, a final limitation to the method is cell-type dependent: cell types that tend to uptake the fluorescent probe or attach to each other will be poorly adapted to the method.

3. DISCUSSION

Fluorescence exclusion measurement (FXm) allows for measurement of volume of single cells over time periods from seconds up to several days. Calibrations of

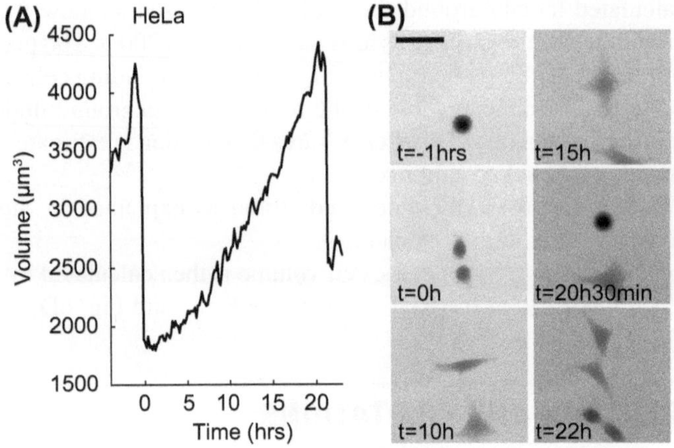

FIGURE 7 Cell volume throughout a complete cell cycle.

(A) Volume trajectory of a HeLa cell. The two volume overshoots at the beginning and the end correspond to transient volume increase in mitosis with the first one corresponding to the mother cell and the second one to the daughter cell. (B) Raw Fluorescence images of the cell in (A) with FXm. Scale bar 50 μm.

volume measurement using FXm compared with geometrically obtained measurement of giant unilamellar vesicles or circular cells (Fig. 5B) yields an accuracy of approximately 10% (Zlotek-Zlotkiewicz et al., 2015). This estimation is however limited by the accuracy of the geometrical measurement itself. Among currently available single-cell volume measurement techniques, the SRM technique is probably the most accurate one with errors estimated to less than 1%, but is limited to suspended cell types (Son et al., 2015). For the 3D reconstruction-based method, accuracy of measurement is limited temporally by the time of acquisition of all the stacks and moreover the precision on volume calculation of such a method has not been demonstrated yet; furthermore, repeated illumination is probably the main limitation for measurement through time due to its toxicity. FXm is however a good compromise between accuracy of measurement and throughput of the experiment as it is the only method that allows parallel measurements on multiple single cells through time.

FXm shows several unique advantages: (1) it is easy to use and adapted to many different cell types, both adherent and suspended; (2) it is largely compatible with classical cell biology tools such as fluorescence imaging or transient drug treatment thus enabling direct investigation of cell volume regulation; (3) it is adapted to timescales ranging from seconds to several days and is therefore suitable for a wide range of biological questions such as short-time response to external perturbations (i.e., osmotic shocks), mitosis, or cell growth through the cell division cycle (see Fig. 7). All these make FXm the most versatile technique currently available to study cell volume regulation.

REFERENCES

Bottier, C., Gabella, C., Vianay, B., Buscemi, L., Sbalzarini, I. F., Meister, J.-J., & Verkhovsky, A. B. (2011). Dynamic measurement of the height and volume of migrating cells by a novel fluorescence microscopy technique. *Lab on a Chip, 11*(22), 3855–3863. http://doi.org/10.1039/c1lc20807a.

Boucrot, E., & Kirchhausen, T. (2008). Mammalian cells change volume during mitosis. *PLoS One, 3*(1), e1477. http://doi.org/10.1371/journal.pone.0001477.

Bryan, A. K., Engler, A., Gulati, A., & Manalis, S. R. (2012). Continuous and long-term volume measurements with a commercial Coulter counter. *PLoS One, 7*(1), e29866. http://doi.org/10.1371/journal.pone.0029866.

Bryan, A. K., Hecht, V. C., Shen, W., Payer, K., Grover, W. H., & Manalis, S. R. (2013). Measuring single cell mass, volume, and density with dual suspended microchannel resonators. *Lab on a Chip, 14*(3), 569–576. http://doi.org/10.1039/c3lc51022k.

Conlon, I., & Raff, M. (2003). Differences in the way a mammalian cell and yeast cells coordinate cell growth and cell-cycle progression. *Journal of Biology, 2*(1), 7.

Cosgrove, D. J. (2005). Growth of the plant cell wall. *Nature Reviews: Molecular Cell Biology, 6*(11), 850–861. http://doi.org/10.1038/nrm1746.

Ernest, N. J., Habela, C. W., & Sontheimer, H. (2008). Cytoplasmic condensation is both necessary and sufficient to induce apoptotic cell death. *Journal of Cell Science, 121*(Pt 3), 290–297. http://doi.org/10.1242/jcs.017343.

Ginzberg, M. B., Kafri, R., & Kirschner, M. (2015). On being the right (cell) size. *Science, 348*(6236), 771–776. http://doi.org/10.1126/science.1245075.

Godin, M., Delgado, F. F., Son, S., Grover, W. H., Bryan, A. K., Tzur, A., … Manalis, S. R. (2010). Using buoyant mass to measure the growth of single cells. *Nature Methods, 7*(5), 387–390. http://doi.org/10.1038/nmeth.1452.

Gray, M. L., Hoffman, R. A., & Hansen, W. P. (1983). A new method for cell volume measurement based on volume exclusion of a fluorescent dye. *Cytometry, 3*(6), 428–434. http://doi.org/10.1002/cyto.990030607.

Gregg, E. C., & Steidley, K. D. (1965). Electrical counting and sizing of mammalian cells in suspension. *Biophysical Journal, 5*(4), 393–405. http://doi.org/10.1016/S0006-3495(65)86724-8.

Habela, C. W., & Sontheimer, H. (2007). Cytoplasmic volume condensation is an integral part of mitosis. *Cell Cycle (Georgetown, Tex.), 6*(13), 1613–1620.

Jiang, H., & Sun, S. X. (2013). Cellular pressure and volume regulation and implications for cell mechanics. *Biophysical Journal, 105*(3), 609–619. http://doi.org/10.1016/j.bpj.2013.06.021.

Mir, M., Wang, Z., Shen, Z., Bednarz, M., Bashir, R., Golding, I., … Popescu, G. (2011). Optical measurement of cycle-dependent cell growth. *Proceedings of the National Academy of Sciences of the United States of America, 108*(32), 13124–13129. http://doi.org/10.1073/pnas.1100506108.

Mitchison, T. J., Charras, G. T., & Mahadevan, L. (2008). Implications of a poroelastic cytoplasm for the dynamics of animal cell shape. *Seminars in Cell and Developmental Biology, 19*(3), 215–223. http://doi.org/10.1016/j.semcdb.2008.01.008.

Park, K., Millet, L. J., Kim, N., Li, H., Jin, X., Popescu, G., … Bashir, R. (2010). Measurement of adherent cell mass and growth. *Proceedings of the National Academy of Sciences of the*

United States of America, 107(48), 20691–20696. http://doi.org/10.1073/pnas.1011365107.

Son, S., Kang, J. H., Oh, S., Kirschner, M. W., Mitchison, T. J., & Manalis, S. (2015). Resonant microchannel volume and mass measurements show that suspended cells swell during mitosis. *The Journal of Cell Biology, 211*(4). http://doi.org/10.1083/jcb.201505058.

Son, S., Tzur, A., Weng, Y., Jorgensen, P., Kim, J., Kirschner, M. W., & Manalis, S. R. (2012). Direct observation of mammalian cell growth and size regulation. *Nature Methods, 9*(9), 910–913. http://doi.org/10.1038/nMeth.2133.

Stroka, K. M., Jiang, H., Chen, S.-H., Tong, Z., Wirtz, D., Sun, S. X., & Konstantopoulos, K. (2014). Water permeation drives tumor cell migration in confined microenvironments. *Cell, 157*(3), 611–623. http://doi.org/10.1016/j.cell.2014.02.052.

Sung, Y., Tzur, A., Oh, S., Choi, W., Li, V., Dasari, R. R., … Kirschner, M. W. (2013). Size homeostasis in adherent cells studied by synthetic phase microscopy. *Proceedings of the National Academy of Sciences of the United States of America, 110*(41), 16687–16692. http://doi.org/10.1073/pnas.1315290110.

Tao, J., & Sun, S. X. (2015). Active biochemical regulation of cell volume and a simple model of cell tension response. *Biophysical Journal, 109*(8), 1541–1550. http://doi.org/10.1016/j.bpj.2015.08.025.

Tzur, A., Kafri, R., LeBleu, V., & Lahav, G. (2009). Cell growth and size homeostasis in proliferating animal cells. *Science*, 167. http://doi.org/10.1126/science.1174294.

Zlotek-Zlotkiewicz, E., Monnier, S., Cappello, G., Le Berre, M., & Piel, M. (2015). Optical volume and mass measurements show that mammalian cells swell during mitosis. *The Journal of Cell Biology, 211*(4), 765–774. http://doi.org/10.1083/jcb.201505056.

CHAPTER

Quantitative methods to study epithelial morphogenesis and polarity

7

B. Aigouy*[,a], C. Collinet*[,a], M. Merkel[§,1,a], A. Sagner[¶,1,a]

*IBDM (Institut de Biologie du Développement de Marseille) UMR 7288, Aix-Marseille Université & CNRS, Marseille, France

[§]Syracuse University, Syracuse, NY, United States

[¶]The Francis Crick Institute, London, United Kingdom

[1]Corresponding authors: E-mail: mmerkel@syr.edu; andreas.sagner@crick.ac.uk

CHAPTER OUTLINE

[a]All authors contributed equally; authorship order is alphabetical.

Methods in Cell Biology, Volume 139, ISSN 0091-679X, http://dx.doi.org/10.1016/bs.mcb.2016.12.004

Abstract

Morphogenesis of an epithelial tissue emerges from the behavior of its constituent cells, including changes in shape, rearrangements, and divisions. In many instances the directionality of these cellular events is controlled by the polarized distribution of specific molecular components. In recent years, our understanding of morphogenesis and polarity highly benefited from advances in genetics, microscopy, and image analysis. They now make it possible to measure cellular dynamics and polarity with unprecedented precision for entire tissues throughout their development.

Here we review recent approaches to visualize and measure cell polarity and tissue morphogenesis. The chapter is organized like an experiment. We first discuss the choice of cell and polarity reporters and describe the use of mosaics to reveal hidden cell polarities or local morphogenetic events. Then, we outline application-specific advantages and disadvantages of different microscopy techniques and image projection algorithms. Next, we present methods to extract cell outlines to measure cell polarity and detect cellular events underlying morphogenesis. Finally, we bridge scales by presenting approaches to quantify the specific contribution of each cellular event to global tissue deformation.

Taken together, we provide an in-depth description of available tools and theoretical concepts to quantitatively study cell polarity and tissue morphogenesis over multiple scales.

INTRODUCTION

Epithelia are the most widespread tissue in the animal kingdom. They are characterized by tight connections between cells called junctions inhibiting diffusion of molecules across the epithelium. Epithelial cells also exhibit an apico-basal polarity underlined by an asymmetric distribution of intracellular and plasma membrane components. This polarized organization allows selective and directed transport of molecules (e.g., nutrients, proteins, ions etc.) across the epithelial barrier.

Most epithelial cells exhibit a second polarity axis, called planar cell polarity (PCP) that lies within the plane of the epithelium. PCP coordinates the alignment of cell external and internal structures locally (between neighboring cells) and globally (across the tissue) (Goodrich & Strutt, 2011). Striking examples of PCP are the coordinated alignment of hairs and cilia found in many epithelia. PCP also coordinates polarized cell rearrangements during convergence and extension movements in vertebrates and invertebrates (Bertet, Sulak, & Lecuit, 2004; Blankenship, Backovic, Sanny, Weitz, & Zallen, 2006; Irvine & Wieschaus, 1994; Tada & Kai, 2012; Walck-Shannon & Hardin, 2014; Wallingford, 2012). Molecularly, PCP depends on the asymmetric segregation of polarity proteins to different sides of a cell (compare Figs. 1A and C). Therefore, in many instances PCP can be defined by a vector, which represents the asymmetric protein localization (Fig. 1A). However, several PCP components (e.g., the atypical cadherin Flamingo (Chae et al., 1999; Usui et al., 1999), Myo-II, E-cad, and Par3 in the germband of *Drosophila* (Bertet et al., 2004; Blankenship et al., 2006; Zallen & Wieschaus, 2004)) localize to both sides of a cell and are therefore characterized by an axial rather than a vectorial polarity (Fig. 1B).

During development, epithelia undergo dramatic morphogenetic changes to acquire their final shape. In most cases these shape changes occur without a loss of epithelial integrity and involve a limited set of cellular events occurring within the plane of the epithelium. Two classes of events can be distinguished, cell deformations and events changing the topology of the cell network (i.e., the local arrangement of cell connections). Cell deformations can be isotropic (i.e., equal changes in all directions), such as reduction or expansion of cell area, or anisotropic such as cell elongation along a specific axis. Most changes in the cell network topology result from three processes: neighbor exchanges, cell divisions, and cell extrusions (Fig. 3A).

A type of neighbor exchange, called T1 transition (Bohn, 2003; Weaire & Hutzler, 2001, p. 246) is defined as the loss of contact between two cells (a, b) that become separated by two adjacent cells (c, d) (Fig. 3A). T1 transitions typically involve four cells and occur in two steps: first the shrinkage of the contact between cells a and b followed by extension of a new contact between cells c and d (Fig. 3A) (Bertet et al., 2004). In some cases more than four cells exchange their neighbors, inducing the formation of rosettes (Blankenship et al., 2006; Nishimura, Honda, & Takeichi, 2012). When T1 transitions occur in an oriented manner throughout an epithelium, e.g., in response to PCP signaling, this results in epithelial convergence and extension. Such polarized cell intercalation has been shown to underlie germband extension in *Drosophila* (Bertet et al., 2004; Blankenship et al., 2006; Irvine & Wieschaus, 1994; Walck-Shannon & Hardin, 2014) and convergent extension in the vertebrate neural tube (Nishimura et al., 2012; Nishimura & Takeichi, 2008), the renal tubules in mouse (Karner et al., 2009) and Xenopus (Lienkamp et al., 2012), and the mammalian cochlea (Chacon-Heszele, Ren, Reynolds, Chi, & Chen, 2012; Wang, Mark, et al., 2005).

Cell divisions increase the number of cells, often contributing to epithelial growth, and modify network topology. Importantly, the orientation of cell divisions has been proposed to bias the orientation of growth in the *Drosophila* wing disc (Baena-Lopez, Baonza, & Garcia-Bellido, 2005). Furthermore, oriented cell divisions have been implicated in releasing tissue level stresses as in the peripheral rim of the *Drosophila* wing discs (Legoff, Rouault, & Lecuit, 2013) and during zebrafish epiboly (Campinho et al., 2013).

During cell extrusion a cell progressively loses contact with its neighbors and leaves the tissue layer. The resulting change in topology is termed T2 transition (Bohn, 2003) (Fig. 3A). Typically, they occur when cells undergo developmentally induced epithelial to mesenchymal transition or die. Cell extrusions by T2 transitions constitute an important mechanism to maintain epithelial homeostasis and are observed in many epithelia, e.g., at the tip of gut microvilli in mammals (Eisenhoffer & Rosenblatt, 2013; Guillot & Lecuit, 2013), the developing fin of zebrafish (Eisenhoffer et al., 2012), and the midline of the developing dorsal thorax in *Drosophila* (Marinari et al., 2012).

In this chapter, we provide an overview of methods to quantify morphogenesis and planar polarity at the cell and tissue scale. We first outline a selection of biological strategies to label cells and unravel polarized protein distribution. We next discuss different microscopy approaches and illustrate in detail suitable methods for image projection, segmentation, and measurement of tissue deformation and polarity. Boxes provide the necessary mathematical background for the quantification methods. At each step, we highlight potential benefits and disadvantages of the presented methods.

1. BIOLOGICAL TOOLBOX

Morphogenesis can be studied with different levels of detail. Below, we first describe biological markers that can be used to monitor coarse-grained tissue deformation, as well as markers to visualize single cell polarity or unravel fine-grained cellular events underlying morphogenesis. We then outline strategies using mosaics to visualize hidden polarities in epithelial tissues.

1.1 CELL AND POLARITY MARKERS

In general, (1) a good biological marker must not perturb the developmental processes studied. Ideally, it should be expressed at endogenous levels from its endogenous locus or be biologically neutral when overexpressed. (2) The marker should be bright, in particular for live imaging. Fluorescent proteins with high quantum yield and low bleaching are preferable. According to our experience, e-GFP, Venus, and td-Tomato perform better than cyan fluorescent proteins. (3) The marker should be as much as possible restricted to and label only the structure of interest (e.g., cell membrane, nucleus, etc.).

Coarse-grained analysis of tissue deformation can be performed by a method called "particle image velocimetry" (PIV) (see Section 3.3) (Raffel, 2007). PIV exploits the natural heterogeneity of the tissue and works with most conventional markers (e.g., cell membrane, nuclei, organelles, etc.).

For a fine-grained analysis of cellular events underlying morphogenesis, an extraction of cell outlines is often required (i.e., segmentation). Optimal reporters for this display a sharp signal around the entire cell perimeter and have low or no cytoplasmic interfering signal (such as vesicles). E-cadherin::e-GFP is such a marker for *Drosophila* tissues. Discontinuous or less-sharp markers can also be used to extract cell outlines (e.g., many planar polarity proteins and components of the cytoskeleton), but with poorer results. Note that to study the distribution of such proteins in the cell, cell outlines may be obtained with a second, more easily segmentable marker (e.g., E-cadherin::e-GFP) and then applied to the signal of the former (planar polarity proteins cytoskeletal components etc.).

1.2 MOSAICS AS A TOOL TO VISUALIZE POLARITY AND MORPHOGENESIS

The limited resolution of conventional light microscopy cannot resolve the relative protein contributions of adjacent cells to their shared boundary. Therefore, a protein localizing only to one side of the cell may appear bipolar (Figs. 1E,F) or nonpolarized (Figs. 1I,K). Hence, vectorial polarity can be hidden (Figs. 1I,K and Section 4.1.2), although in many situations, an axial polarity can still be detected (Figs. 1D–F).

Genetic mosaics, i.e., tissues consisting of two or more genetically distinct cell populations, provide an elegant solution to this problem, because they allow the expression of a labeled protein in only a subset of cells (Figs. 1G,H,L,M). As a consequence, any labeled protein localized at the interface between labeled and unlabeled cells comes only from the labeled cells. Thus, this interface can be used for the quantification of magnitude and direction of PCP vectors (Figs. 1D–M and Section 4.1.1). Note however that there may be rare cases where vectorial polarity can be measured without requiring mosaics, for example, when few cells show higher fluorescence intensity than their neighbors (Besson et al., 2015).

Mosaics can be induced in several ways in several organisms. They can be created by injecting, infecting, or transfecting cells with vectors or by creating chimeras between different animals (Kretzschmar & Watt, 2012). However, the two most widely used methods are the Flp/FRT and the Cre/lox systems (Nagy, 2000; Sauer & Henderson, 1988; Xu & Rubin, 1993). In both cases, the expression of a recombinase (Flippase/Flp or Cre) induces recombination between two target DNA sequences (FRT or loxP sites). Furthermore, both systems are highly versatile as they allow tight regulation of recombinase activity in time and space (Danielian, Muccino, Rowitch, Michael, & McMahon, 1998; Hayashi & McMahon, 2002; Xu & Rubin, 1993). Recombination in cis (i.e., between sites on a single DNA molecule) or in trans (on different molecules) can be mediated using both systems. When two recombination sites flank a DNA sequence, recombination in cis results in the excision of the flanked DNA fragment. This is often used to delete parts of a gene to mutate it or to excise a central "Stop" cassette to conditionally activate gene expression (Kretzschmar & Watt, 2012; Srinivas et al., 2001). This for example allows the generation of multiple small labeled clones in an otherwise unlabeled tissue. In *Drosophila*, recombination in trans between two FRT sites is used to exchange distal arms of homologous chromosomes during mitosis (Golic, 1991; Xu & Rubin, 1993). Such a recombination can generate two genetically distinct daughter cells from a heterozygous mother cell.

Genetic mosaics can also be used to label clonally related cells to visualize local morphogenesis. Indeed, clone shape results from all the cellular events occurring

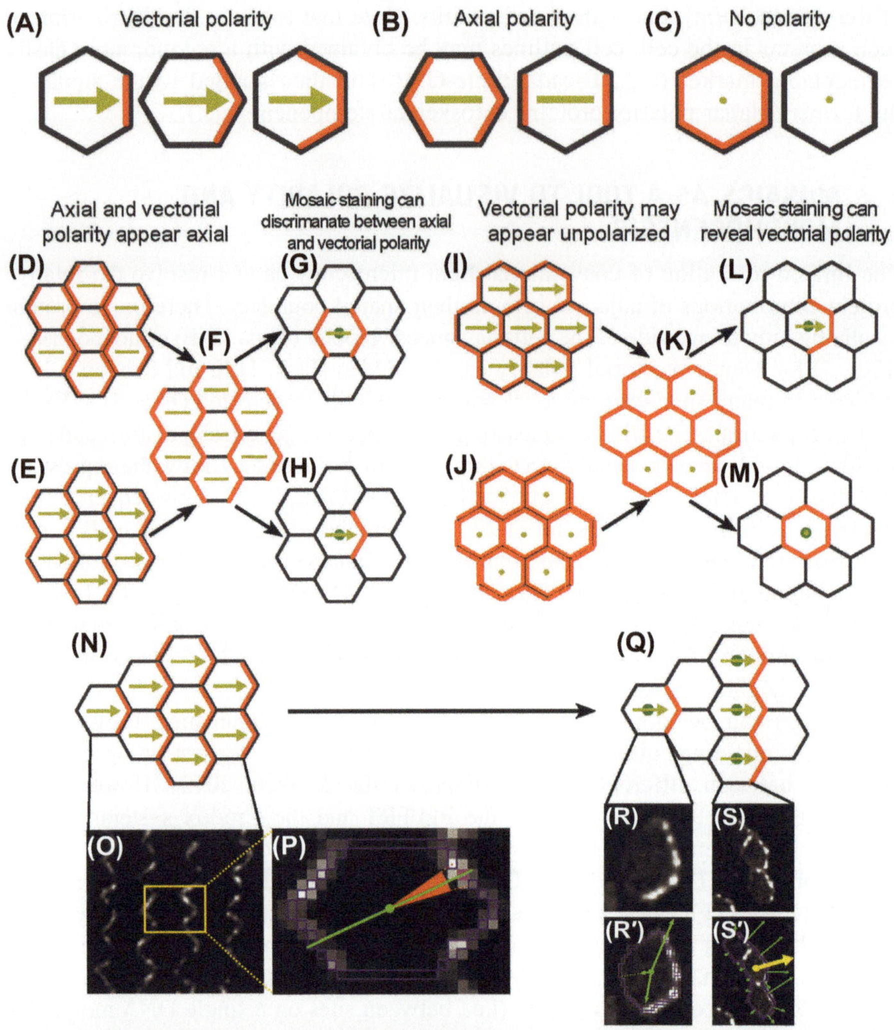

FIGURE 1 Axial and vectorial polarities in cells and tissues.

(A) Three examples showing an asymmetric distribution of a membrane-associated polarity protein (red) within a cell. This kind of polarity is visually represented by an arrow and can be quantified by a vector. Throughout the figure, cell membrane is shown in gray. (B) Two examples of axial polarity for a single cell. Axial polarity is characterized by the presence of same amounts of a polarity protein on opposing sides of a cell. Magnitude and orientation of axial polarity is represented by a bar and can be quantified by a "nematic." (C) In the absence of protein asymmetries, neither vectorial nor axial polarity can be detected (*yellow dot*). (D–M) Tissues with axial and vectorial polarities and how they appear in microscopy images (F–H, K–M). (D) Tissue composed of axially polarized cells (left cell in (B)). (E) Tissue composed of vectorially polarized cells (middle cell in (A)). (F) Vectorial and axial polarity within the clone. For example, elongated clone shape indicates anisotropic cell behaviors (e.g., oriented cell divisions, oriented neighbor exchanges, and/or cell elongation), while round clones may result from either absence of anisotropies or opposing anisotropic cellular behaviors. Similarly, clone fragmentation indicates a high rate of neighbor exchange. However, absence of clone fragmentation does not necessarily indicate absence of neighbor exchanges. This is the case in the *Drosophila* wing, where clonally related cells mostly remain attached to each other despite extensive cell rearrangements (Aigouy et al., 2010; Heller et al., 2016). Additionally, reproducible differences in clone size indicate heterogeneous growth within the tissue. Also, obtaining few snapshots of the development of clones allows to observe their deformation as well as to infer global tissue deformation by triangulation between them (Heemskerk, Lecuit, & LeGoff, 2014).

2. IMAGING TOOLBOX

Imaging (i.e., the acquisition of images with a microscope) is an essential technique to understand the cell biological and molecular basis of development. Light microscopy allows to observe tissues with cellular and subcellular resolution. Since

appear axial in microscopy images. (G, H) Mosaic tissue containing cells expressing a tagged planar cell polarity (PCP) protein (*green dot*) surrounded by unlabeled cells. Mosaics allow to discriminate between axial (G) and vectorial (H) polarities. Also, the direction of vectorial polarity is made evident (H). (I) Tissue composed of vectorially polarized cells (right cell in (A)). (J) Tissue composed of nonpolarized cells (left cell in (C)). (K) Vectorially polarized and nonpolarized cells both appear unpolarized in microscopy images. (L, M) Mosaics can discriminate between vectorially polarized cells (L) and nonpolarized cells (M). (N–S′) Axial and vectorial polarity measurements using mosaics. (N, O) PCP organization in the pupal wing of *Drosophila*. Note that the cells exhibit an axial polarity. (P) Axial cell polarity is quantified by a nematic, which is the sum of individual pixel nematics (*green bar*) that pass through the cell center (*green circle*) and the respective pixel (magenta). The magnitude of each pixel nematic is given by the pixel intensity times the angular distance of neighboring pixels (red). (Q) Mosaic tissue containing cells expressing tagged PCP proteins (*green dots*). (R) Isolated cell expressing a tagged polarity protein surrounded by unlabeled cells. (R′) Corresponding vectorial polarity measurement for an isolated cell, as described in (Besson et al., 2015). The cell polarity vector is the sum of polarity vectors for the individual pixels of the cell outline (*green arrows*), which point from the cell center (*green dot*) to the respective pixel. The length of each pixel vector is the pixel intensity times the distance between cell center and pixel. (S) Clonal population composed of three cells expressing a tagged PCP protein, surrounded by unlabeled cells. (S′) Corresponding vectorial polarity measurement for this 3-cell clone (Sagner et al., 2012). The polarity vector of the clone (yellow) is the sum of polarity vectors (*green arrows*) computed for all interfaces (*blue lines*) that outline the clone. The vector for a given interface is perpendicular to the interface pointing outside the clone, and the vector length corresponds to the sum of pixel intensities along the segmented interface.

morphogenesis is a dynamic process that is better understood when followed in real time, live imaging is the primary approach to study morphogenesis in many labs. Furthermore, although cell and tissue polarity can be studied in fixed tissues, understanding how polarity is established and maintained in face of dynamic cell behaviors greatly benefits from live imaging.

A microscopy image is nothing but a visual representation of a two-dimensional (x,y) matrix of intensity values (Fig. 2A). Each element of this matrix is called a pixel. Pixels can be coded over *n* number of bits, typically 8–16 for most biological images. Pixel intensities in 8-bit images therefore range from 0 to 255 (i.e., 2^8 gray levels) and in 16-bit images from 0 to 65,535 (i.e., 2^{16} gray levels). Imaging datasets of labeled biological samples can have more than just two-dimensions when also the sample depth (z-dimension), time (t-dimension), and multiple colors (c-dimension) are considered. In this section, we present the most common types of microscopy and discuss their applications, advantages, and disadvantages relatively to different biological questions (Table 1).

2.1 TYPES OF MICROSCOPY

The simplest and cheapest microscopy setup is widefield imaging (also called epifluorescence). An epifluorescence microscope captures the light emitted by the whole sample with an objective lens and records each image at once with a CCD camera. Widefield imaging is therefore fast, but it also collects out-of-focus light from regions above and below the focal plane. While this is generally not a problem when imaging thin samples (e.g., cells in a petri dish), the out-of-focus light reduces the contrast in images of whole mount organisms.

This limitation is overcome in confocal microscopy, which allows to acquire thin optical sections to obtain sharp 3-D representations of the sample. Laser scanning confocal microscopy (LSCM) and spinning disk confocal microscopy (SDCM) are the two most commonly used types of confocal systems. They both utilize pinholes, i.e., apertures in the detection path that reject out-of-focus light generating

Table 1 Comparison of Different Types of Microscopy

Microscope/ type of analysis	Polarity	Tissue scale morphogenesis	Cell scale morphogenesis	Phototoxicity[a]
Widefield	+	+++	+	+
Confocal	+++	+	+++	++
Spinning disk	+++	+++	+++	+
Light sheet microscope	+	+++	+++	+
Two-photon microscopy	+++	+	+++	+++

[a] *more "+" means more phototoxic.*

images with higher contrast compared to widefield. LSCM and SDCM differ in how the sample is scanned and signals are detected, making the two systems optimal for different applications.

In LSCM a laser is moved with mirrors across the sample, and the image is generated pixel-by-pixel by recording the emitted photons using a photon-multiplying tube (PMT). The PMT converts the photons into electrical/analog signals stored as intensity values in an image by the recording software. LSCM offers good x-, y-, and z-resolution close to the theoretical limit of conventional light microscopy and a high signal-to-noise ratio (S/N). Properly adjusting the scan speed parameter, i.e., the time the laser spends on each pixel (also referred to as pixel-dwell time), allows to improve S/N and therefore the quality of the acquired images. Traditionally, the main drawback of LSCM is the overall low speed of acquisition, especially when multiple channels, large (x, y) fields and big z-volumes need to be recorded. Nevertheless, modern LSCM systems can be used to perform relatively fast live imaging. Alternatively, scan speed can be increased by reducing the field of view or at the expense of S/N. Taken together, LSCM is an excellent choice for acquiring images with high resolution and good S/N, when slow acquisition is not a problem. Indeed, LSCM is most often the prime choice for fixed samples stained by immunofluorescence.

A recent development in LSCM is the use of multiphoton excitation (most frequently two-photon on conventional systems) (Denk, Strickler, & Webb, 1990). Classically, fluorophores are excited by a single photon with a wavelength shorter than that of the emitted photon. In contrast, with multiphoton excitation, the fluorophore is excited by simultaneous absorption of two or more photons with a larger wavelength (multiphoton effect). This offers three main advantages compared to classical LSCM. First, photons with longer wavelengths can penetrate further into the sample and thereby allow deeper imaging (Helmchen & Denk, 2005). Second, since the photon density required for the multiphoton effect is reached only at the focal plane, all emitted photons are "in-focus" and there is no need for pinholes to filter out-of-focus light. This dramatically increases the S/N giving this setup an advantage when imaging light scattering samples (Denk & Svoboda, 1997). Third, because fluorophores are only excited in the focal plane and not, as in classical LSCM, throughout the sample, multiphoton microscopy causes less photobleaching, especially when acquiring large z-volumes (So, Dong, Masters, & Berland, 2000). However, multiphoton imaging also has disadvantages. First, since the scanning principle is similar to regular LSCM, multiphoton imaging is slow. Additionally, it is not well suited for multicolor imaging because multiphoton lasers usually need time-consuming steps of tuning to switch between different wavelengths and cannot be easily combined with visible lasers for regular LSCM. Finally, the high photon density of multiphoton lasers can cause photodamage and in many cases is not well tolerated by the sample. In conclusion, multiphoton microscopy provides a good alternative to regular LSCM when deep sample penetration is required.

In SDCM, instead of using a single pinhole, a rotating disk with several thousand pinholes (also called Nipkow disk) is put into the conjugate image plane (Conchello & Lichtman, 2005). In 2002, the Yokogawa spinning disc unit was developed, this SDCM device focuses a single illumination beam on the spinning disk, effectively multiplying it into thousands of parallel beams focused onto the sample by the objective lens (Inoué & Inoué, 2002; Tanaami et al., 2002). Upon disk rotation the parallel beams simultaneously scan the entire specimen at a speed that is much higher than LSCM. The emitted light coming from the sample is collected back through the pinholes producing an optical section that is registered at once with an image by CCD cameras. Such a parallel scanning and reading strategy requires low sample illumination and thereby reduces photobleaching and photodamage. Thus, SDCM is a good choice to image living samples and fast developmental processes (Conchello & Lichtman, 2005; Tanaami et al., 2002; Wang, Babbey, & Dunn, 2005). Also this is an excellent device to image large samples with cellular resolution using image stitching (see Section 3.2) (e.g., Bosveld et al., 2012; Etournay et al., 2015). Disadvantages of this setup are that the size of the pinholes in the spinning disk is fixed and usually adjusted to yield optimal results with a certain objective (usually 40× or 100x). Thus, when imaging with other magnifications, the size of the pinholes in the disk does not provide an optimal photon yield and z-resolution. In summary, spinning disk microscopes are a good choice for imaging fast developmental processes and/or entire tissues with cellular resolution.

Recently, selective plane illumination microscopy (SPIM) has emerged as an interesting technique to study developmental processes. In SPIM, the specimen is illuminated with a thin light sheet, and the emitted photons are collected in perpendicular direction (Keller & Stelzer, 2008; Weber & Huisken, 2011). SPIM has two main advantages. First, optical sectioning is produced by illuminating only the desired planes rather than rejecting out-of-focus light with pinholes. This dramatically reduces photobleaching and photodamage allowing long-term imaging of living samples. Second, since the entire (x, y) plane is illuminated at once, this technique allows fast 3-D scanning of the specimen with sensitive high-speed CCD cameras. By embedding the sample in a rigid transparent medium (e.g., agarose), it is possible to rotate it and collect 3-D stacks at different angles, providing isotropic resolution in the x-, y-, and z-dimension and allowing in toto 3-D imaging of living specimens (Huisken & Stainier, 2009; Weber & Huisken, 2011). This technique may however produce light scattering artifacts in opaque specimens (e.g., shading due to opaque objects). Recent development including dual-side illumination with multidirectional SPIM (mSPIM) have been shown to significantly alleviate these problems (Huisken & Stainier, 2007; Krzic, Gunther, Saunders, Streichan, & Hufnagel, 2012; de Medeiros et al., 2015). Finally, since SPIM can produce large datasets, 3-D reconstruction can be long and computationally intensive. In conclusion, SPIM is currently limited by difficulties in data handling and storing but remains a promising technique for fast three-dimensional and multiview imaging of biological samples with very low photobleaching.

2.2 TIPS AND PITFALLS TO OPTIMIZE IMAGING PROTOCOLS

Microscopy images have to respect certain criteria to provide meaningful data quantifications. In this section, we discuss potential tips and pitfalls to optimize imaging protocols and acquire images appropriate for image analysis.

2.2.1 *Optimizing the dynamic range, the signal-to-noise ratio, and the speed of imaging*

A good image is an image that (1) uses the full dynamic range provided by the bit depth (8–16 bit), (2) in which pixels neither reach the maximum nor the minimum value of such range, (3) has a good S/N ratio. In fluorescence microscopy, this can be achieved by increasing the power of the illuminating laser. This generates higher signals and improves the S/N ratio without compromising on imaging speed. The major downside of using high laser power is photobleaching and photodamage (Section 2.2.2). Usually, appropriate illumination power settings are those that maximize S/N while minimizing photobleaching.

When increasing laser power is detrimental for the sample, other parameters controlling photon detection or photon conversion into digital signals can be tuned to improve image quality. In epifluorescent, SDCM, and SPIM the amount of collected signal depends on the camera exposure time (i.e., the total time the camera spends collecting photons). Increasing exposure time yields brighter images with better S/N at the expense of imaging speed. It is important, especially for live imaging, to find the right trade-off between exposure time and image quality. In LSCM, detection can be regulated by controlling the pixel dwell time (equivalent of exposure time) or the pinhole aperture. S/N can be increased by either decreasing the scan speed (i.e., increasing the dwell time) or by averaging multiple scans. Of note, in LSCM, following an appropriate phase correction, dual scanning (scans of sequential lines in a left-right-left manner) can be activated to reduce the scan time while keeping dwell time constant. This option is especially useful for live cell imaging. Furthermore, when the emitted fluorescence spectra do not overlap, simultaneous detection of multiple fluorophores can be used to avoid compromising on acquisition speed. Finally, differently from SDCM, in LSCM the aperture of the pinholes can be adjusted allowing to increase the thickness of the optical slice (i.e., decreases confocality) thereby collecting more photons without decreasing imaging speed.

The gain parameter controls the electron amplification process. Upon increase, higher signals are obtained from the same amount of photons collected. This allows a better use of the dynamic range but also introduces more noise, having a moderate effect on image quality.

2.2.2 *Preventing photobleaching and photodamage*

Common problems in live imaging are photobleaching and sample photodamage. Photobleaching is the gradual loss of emitted fluorescence when labeled samples are illuminated with a laser. This is due to the photochemical alteration of fluorophores resulting in the permanent loss of the ability to emit photons. Photobleaching is one of the first signs that a sample is either illuminated for too long or with too high laser intensity. Although photobleaching can be corrected a posteriori, it is preferable to reduce the intensity of the exciting laser or exposure of the sample to light whenever photobleaching is detected.

Light can also damage the sample due to the generation of toxic side products (usually reactive oxygen species). Photodamage can lead to slower or abnormal development. Comparisons to fixed or living samples not subjected to the imaging protocol at comparable stages of development are usually sufficient to detect photodamage. Photodamage is therefore an important factor to be taken into consideration when designing live imaging protocols, especially when studying long morphogenetic processes.

Altogether, the choice of the imaging approach depends on the biological system and the question addressed, and proper tuning of imaging parameters are key to obtain high-quality images for analysis while preserving sample health.

3. IMAGE PROCESSING TOOLBOX

In this toolbox, we present tools to project 3-D stacks into 2-D images and create large views of tissues by combining small high-resolution views. Next we describe algorithms to follow and measure cell and tissue movements with or without segmentation.

3.1 PROJECTION ALGORITHMS

Although often imaged with 3-D stacks, epithelia are well approximated in 2-D. Hence, image analysis is usually performed in 2-D images. Such images are obtained by converting the three-dimensional information contained in a (x, y, z) stack into a single two-dimensional (x, y) image, through a process called a projection. Numerous projection algorithms are available, but according to our experience, some are better suited than others for segmentation and polarity quantification (Table 2).

Table 2 Comparison of Different 2-D Projection Algorithms

Projection algorithms/ experiments	Polarity	Tissue scale morphogenesis	Cell scale morphogenesis
Maximum	++	++	++
Average	+++	+++	+
Stack focuser	+	+	+++
Merkel et al. (2014)	++	+	+++
Extended depth of field	+	+	++

More "+" means better.

The most widespread projection algorithm is the maximum (Max) projection. It is a pixel-based algorithm that identifies the pixel at position (x_p, y_p) with the maximum intensity value along the Z-axis of the stack (Figs. 2B,D). This brightest pixel is then used in the 2-D projection image. Max projection works particularly well when the signal is spatially restricted, especially in the z-direction, as for most junctional markers. However, by construction Max projection gathers noise from every focal plane, often yielding a noisy image when the input is noisy.

Another useful pixel-based algorithm is the average projection. This algorithm computes for every pixel at position (x_p, y_p) the average pixel intensity over all z-planes (Figs. 2B,D). This average is then assigned to the pixel at position (x_p, y_p) in the 2-D projection (Figs. 2B,D). In contrast to Max projection, the average projection does not discard any signal. However, average projections often show poorer contrast than other algorithms, especially for large z-stacks when the signal of interest is confined to few planes. Although average projection is not the best choice for segmentation, it remains the preferred algorithm for polarity quantification.

Often, Max projection is sufficient to easily segment images. However, when this is not the case, we recommend using one of the "focusing" projection algorithms described below. These algorithms assign to each region of the projection the signal from the most focused z-plane at this position. This usually enhances image contrast and hence simplifies image segmentation. The extraction of in-focus signals can be done pixel-by-pixel, using the "extended depth of field" ImageJ/FIJI plugin (Fig. 2D) (Forster, Van De Ville, Berent, Sage, & Unser, 2004; Preibisch, Saalfeld, & Tomancak, 2009; Schindelin et al., 2012; Schneider, Rasband, & Eliceiri, 2012), or for image blocks of defined width and height (Fig. 2C). "Stack focuser" is an efficient block-based algorithm that runs an edge detection filter on each z-plane of the block. The z-plane presenting the sharpest edges is used for the 2-D projection. The major drawback of this algorithm is that the sharpest signal is sometimes not the signal of interest. This can give rise to projection artifacts, where neighboring blocks carry signal coming from different depths with an intensity mismatch (Fig. 2D′). Such block artifacts can be corrected by ensuring that the selected z-planes of adjacent blocks have similar depths. This correction is usually achieved using surface smoothing algorithms (Heller et al., 2016; Legoff et al., 2013; Merkel et al., 2014). This correction can be combined with a Max projection of a few planes surrounding the best-focused plane to further attenuate block artifacts (Merkel et al., 2014). Of note, block-based projections are by construction sensitive to sample curvature and can give different results depending on chosen block size. Consequently, block-based projections are not optimal for polarity measurements.

Altogether, some projection algorithms are optimized for polarity measurements (e.g., average projection) while others are better suited for segmentation (e.g. Max projection). Note, however that images can first be projected with one algorithm and the resulting segmentation mask (see Section 3.4) can be applied to a different projection of the same image for measurement (see Section 4).

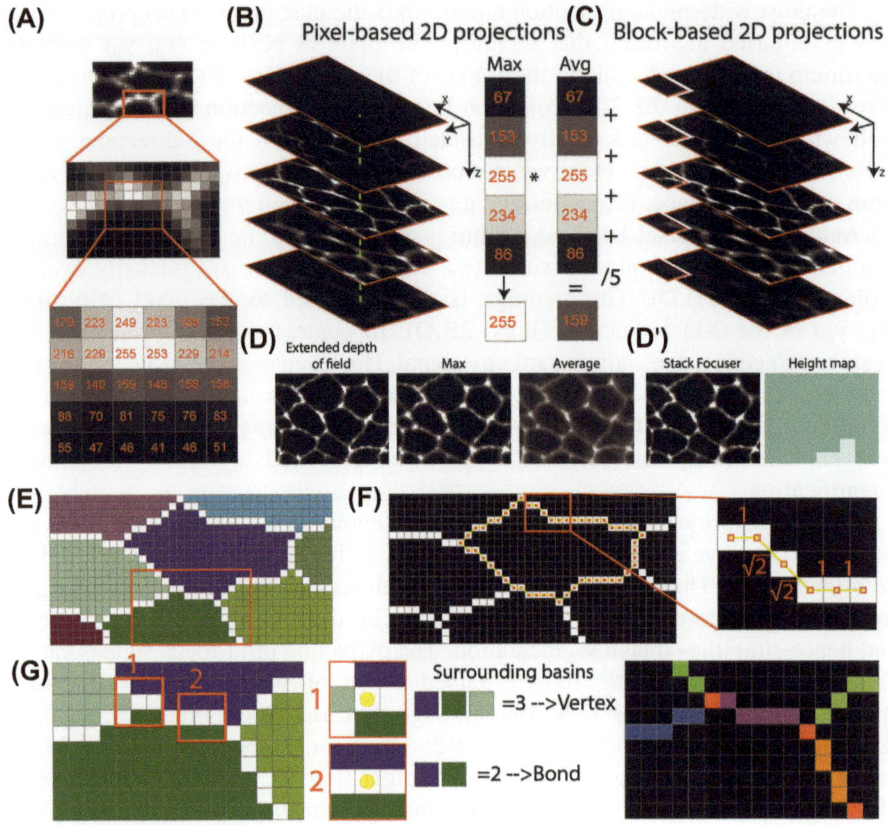

FIGURE 2 Projection algorithms and cell segmentation.

(A) (Top) Image of an epithelial tissue. (Middle) Blow up of the same image, note that an image is composed of picture elements (pixels). (Bottom) Images are nothing but a matrix of pixel intensities (*red numbers*). (B) Examples of pixel-based 2-D projection algorithms. (Left) Series of two-dimensional (x,y) optical sections (along the z-axis) of an epithelium. (Middle) Max projection of the pixels along the *dashed green line*. The Max projection algorithm finds the pixel with the maximum intensity along the z-axis (asterisk). This pixel is used in the 2-D projection (arrow). (Right) Average (Avg) projection of the pixels along the *dashed green line*. The average pixel intensity is used in the 2-D projection. (C) Example of a block-based 2-D projection algorithm. The stack is cut into a series of blocks. Each block spans the whole z-axis and has defined width and height. One such block, located on the top left side of the stack, is indicated. In most block-based projections, only the most informative z-plane of the block is kept for the 2-D projection (not shown). (D) Results of the extended depth of field (left), maximum (middle), and average (right) 2-D projections of the stack shown in (B). (D′) Result of the Stack Focuser projection (left). Corresponding stack focuser height map indicating the depth of the z-plane used for 2-D projection. White blocks originate from deeper z-planes than blue ones. (E) Watershed segmentation of the cell shown in (A). Each basin, i.e., cell cytoplasm, is attributed a unique color. Cell outlines

3.2 IMAGING LARGE SAMPLES WITH CELLULAR RESOLUTION/ STITCHING

Studying morphogenesis or polarity greatly benefits from imaging with cellular resolution (see Section 4). However, high resolution is obtained at the expense of the size of the field of view. Hence, with conventional microscopes it is not possible to capture an entire tissue with cellular resolution. This drawback can be overcome by moving the sample to acquire a mosaic of several overlapping x-, y-, z-stacks (i.e., tiles) that together cover the entire tissue. Bringing those tiles together into a single image of the tissue is called stitching. In practice, stitching algorithms identify features within stacks used as landmarks. Matching these landmarks between adjacent tiles allows to compute the x, y, z translation required for their alignment. Some rules need to be observed to obtain efficient stitching: (1) Neighboring tiles must slightly overlap (according to our experience, 5–10% overlap is sufficient) and (2) The time difference between the acquisitions of overlapping tiles should be sufficiently small to preserve landmark correspondence. We have obtained best results using the highly flexible "Grid/Collection stitching" FIJI plugin (Preibisch et al., 2009).

3.3 PARTICLE IMAGE VELOCIMETRY

PIV allows to visualize the relative speed of displacement of different regions in a tissue and can be used to measure tissue deformation (see also Section 4.2.1). PIV was initially designed to study fluid dynamics by adding small (nonperturbing) tracer particles to the fluid (Raffel, 2007). Following the displacement of the particles allows to characterize the fluid's flow field. This technique can be directly applied to study subcellular, cellular, or tissue motion upon injection of fluorescent beads or by using the natural heterogeneities of fluorescently labeled images (Aigouy et al., 2010; Collinet, Rauzi, Lenne, & Lecuit, 2015; Ganguly, Williams, Palacios, & Goldstein, 2012; He, Doubrovinski, Polyakov, & Wieschaus, 2014; Levayer & Lecuit, 2013; Mayer, Depken, Bois, Julicher, & Grill, 2010;

are indicated by white pixels. (F) Pixels of the perimeter of the central cell in (E) are outlined (left). Perimeter length is quantified as the sum of the lengths of the (yellow) lines connecting the centers of adjacent pixels (*red squares*). (Right) Blow up of the region indicated by a rectangle on the left image. A vertical or horizontal connection is 1 pixel long and an oblique one line has a length equal to $\sqrt{2}$, i.e., the diagonal of a pixel. (G) (Left) Blow up of the region indicated by a rectangle in (E). (Middle) Blow ups of two 3×3 squares located at two positions along the outline of the *dark blue cell* (1 and 2). Counting the different basin colors surrounding the central pixel (indicated by a *yellow circle*) of a square allows to determine if a pixel belongs to a boundary (i.e., is surrounded by exactly two different basins) or to a vertex (i.e., is surrounded by three or more basins). (Right) Image with segmented vertices indicated in red, cell boundary pixels labeled with random colors and cytoplasmic pixels indicated in black.

Merkel et al., 2014; Supatto et al., 2005). PIV divides the image into rectangular blocks and tries to infer, how much and in which direction each of these blocks displace from one frame to the next. As a measure of block similarity, spatial cross-correlations between pixel intensities are computed across consecutive frames. Knowing the displacement of all blocks over time allows to calculate the flow field of the sample. PIV works for a broad range of tissues irrespective of the staining used (e.g., cell membranes, nuclei, intracellular organelles, etc.) and is fast because it does not require segmentation. Several PIV implementations are available in ImageJ/FIJI and Matlab (i.e., MatPIV by J. K. Sveen).

3.4 SEGMENTATION

PIV provides rapid quantification of tissue movements, but is not sufficient to deeply characterize the cellular events at the basis of morphogenesis. To access this information, images need to be segmented. Here, we focus on algorithms to segment membrane-labeled cells.

Segmentation is nothing but a simplification process. For our purpose, segmentation means assigning image pixels to two categories, (1) the boundary pixels and (2) the cytoplasmic pixels. The watershed segmentation is often used for that (Vincent & Soille, 1991). In the watershed representation, the 2-D image is seen as a landscape; regions with high pixel intensities (in our case, cell membranes) are mountains, and regions with low pixel intensities (in our case, cell cytoplasm) are valleys. Immersing this virtual landscape in water will create one basin per valley. Each basin is assigned a unique identity represented by a color. Raising the water level expands individual basins until adjacent basins meet. When this occurs, a so-called watershed line is drawn between them. This watershed line labels the cell outline while the colored basin corresponds to the cell cytoplasm (Fig. 2E). There are several implementations of the watershed algorithm available for ImageJ and Matlab. In FIJI/ImageJ, we recommend using the "graylevel watershed" plugin (http://bigwww.epfl.ch/sage/soft/watershed/) rather than the default implementation. Blurring images is often useful before using the watershed algorithm, as noise leads to oversegmenation (i.e., too many cells detected). However, to preserve cell outlines it is important to use so-called "edge preserving filters" (such as Gaussian, bilateral, anisotropic diffusion, or Kuwahara filter) (Pal, Chakrabarti, & Ghosh, 2015).

Many cellular properties can be directly extracted from the watershed segmentation mask. Cell area, for example, corresponds to the number of cytoplasmic pixels of a cell plus half of the number of perimeter pixels (since cell outline is shared with adjacent cells). The length of the cell perimeter (in pixels) is obtained by summing the distances between centers of adjacent cell outline pixels (Fig. 2F). A vertical or horizontal connection is 1 pixel long, and an oblique connection has a length equal to $\sqrt{2}$ (Fig. 2F). Cell perimeter or area in pixels can be converted to real units by multiplying them by the pixel size or the square of the pixel size in real units, respectively. The x- and y-coordinates of the cell center are obtained by averaging the x- and y-coordinates of all the pixels of the cell cytoplasm, respectively. Cell elongation, i.e., cell shape anisotropy, has been quantified in different ways (Aigouy et al., 2010; Blanchard et al., 2009; Brodland, Chen, & Veldhuis, 2006; Etournay et al., 2015; Graner, Dollet, Raufaste, & Marmottant, 2008; Guirao et al., 2015; Lynch, Veldhuis, Brodland, & Hutson, 2014). One widespread method first fits an ellipse to the cell outline or cytoplasm pixels. Then, cell elongation is defined as the ratio between the lengths of the long and short axes of the ellipse. The orientation of cell elongation is given by the orientation of the long axis of the ellipse. Note however that there are different ways to fit an ellipse to a set of pixels, which might give slightly different elongation and orientation (Stojmenovic & Nayak, 2007).

To extract topological properties such as neighbor number or to identify individual cell–cell contacts, further segmentation of the watershed mask is required. To identify vertices (i.e., the point of contact between three or more cells) and bonds (i.e., points of contact between exactly two cells), a 3×3 pixel mask needs to be slid over every pixel of the cell outline. If the central pixel is surrounded by exactly two-colored basins, then the pixel belongs to a boundary (Fig. 2G). In contrast, if this pixel is surrounded by three (or more) different basins then it is a vertex (Fig. 2G). The length of these newly identified cell–cell contacts can be measured as previously described for the perimeter, the first and the last pixels of the contact being the vertices. Finally, cell neighbor number can be obtained by counting the number of vertices along the cell perimeter.

Further information can be extracted by tracking cells. Tracking cells in a time-lapse movie means assigning unique identities to cells in the field of view and consistently reidentifying these cells over subsequent frames. Reidentification is usually based on cell position, cell neighborhood, and/or on cellular features such as cell area or shape. Dividing and extruding cells can be found by analyzing appearance or disappearance of cell tracks. For example, cell division can be detected as the sudden appearance of two new cells whose combined area is similar to that of the mother cell prior to division. A T2 transition is detected as a cell disappearing from the field of view. Note that in all cases, cells entering or leaving the field of view from the sides should not be considered as dividing or dying. A T1 transition is defined as a neighbor exchange involving four adjacent cells (Fig. 3A). To detect occurring T1 events, it is necessary to identify for each cell–cell junction in the movie the pairs of cells sharing (a and b) or separated (c and d) by this junction (Fig. 3A). When in a subsequent frame, cells a and b are separated by a junction between cells c and d, a T1 transition occurred (Fig. 3A). A rosette is an exchange of neighbors between five or more cells. It is characterized by the concomitant shrinkage of two or more adjacent contacts to a single point prior to the resolution of the rosette. The simplest way to detect a rosette is to detect two or more adjacent contacts with a similar orientation having a combined length inferior to a user-defined threshold (Blankenship et al., 2006).

Note that several free software suites are available to segment and analyze epithelial cells. They conveniently combine in a single-tool segmentation algorithms, cell feature extraction and, in some case, even allow to track cells (Aigouy et al., 2010; Cilla et al., 2015; Etournay et al., 2016; Gelbart et al., 2012;

Heller et al., 2016; Khan, Wang, Wieschaus, & Kaschube, 2014; Leung & Fernandez-Gonzalez, 2015; Mosaliganti, Noche, Xiong, Swinburne, & Megason, 2012; Sagner et al., 2012).

4. MEASUREMENTS

4.1 PLANAR POLARITY

In this section, we discuss different segmentation-based methods to quantify vectorial and axial polarity. In general, methods based on segmentation are more precise, but also more labor intensive.

4.1.1 Vectorial polarity

For light microscopy images, membrane-associated vectorial polarity should be measured on mosaic stainings (see Section 1.2). Several approaches have been used to quantify polarity vectors for a connected group of labeled cells in a mosaic tissue (Ambegaonkar & Irvine, 2015; Brittle, Thomas, & Strutt, 2012; Devenport, Oristian, Heller, & Fuchs, 2011; Merkel et al., 2014; Sagner et al., 2012), which for the sake of simplicity we call "clone" throughout this section. Here, we present two methods to measure polarity of single cells and one method to measure the polarity of "clones."

Besson et al. (2015) define the polarity vector of a single cell from the angular distribution of pixel intensities along the cell outline (Fig. 1R′). The polarity vector of the cell is defined as the sum of polarity vectors for pixels along the cell outline. Each pixel vector is computed as an outward directed vector, which points from the cell center to the pixel itself, and has a length equal to the product between the pixel intensity and its distance from the cell center. To include most of the signal at cell–cell junctions, the segmentation mask can be expanded in width. Besson et al. normalize the polarity vector of a cell by the product between the total cell outline intensity and an average cell radius. This allows to compare cell polarity vectors among different cells.

By construction, this method is sensitive to cell shape. In particular, a cell with a homogeneous signal along its outline may appear polarized when asymmetrically shaped. To overcome this shape dependency, we propose a modified version of the method described above. In this version, the length of each pixel vector is the product between the pixel intensity and the angle interval between its two neighbors along the outline (Section 4.1.2 and Fig. 1P). Also, we propose to normalize by the sum of all vector lengths. As a consequence, a cell with constant pixel intensity along its outline will have a polarity vector of zero.

To measure the polarity vector of "clones" of cells, Sagner et al. (2012) use the segmented "clone" outline. The "clone" polarity vector (thick-yellow arrow) is the sum of polarity vectors (green arrows) of each cell–cell interface along the "clone" outline (Figs. 1Q,S,S′). Each interface polarity vector is an outward pointing vector with a direction perpendicular to the interface (defined by the straight line connecting its two vertices, blue lines in Fig 1S′) and a length equal to the sum of pixel intensities along the segmentation mask. To compare samples with different overall intensities, the "clone" polarity vector can be normalized by the total outline intensity. As a result, the length of the normalized "clone" polarity vector varies between zero (i.e., the "clone" is not polarized) and one (i.e., all nonzero interface vectors point in the same direction).

Finally, in cases where polarity markers are clearly distinguishable from the cell membrane, neither clonal staining nor segmentation is needed to determine vectorial cell polarity. In such a case, the location of the signal of interest with respect to a reference signal (for instance the cell nucleus or the cell membrane) can be quantified using so-called "fluorescence cross-correlations" (Matis, Axelrod, & Galic, 2012).

4.1.2 Axial polarity

Several methods to quantify axial polarity are available (Aigouy et al., 2010; Bardet et al., 2013; Bosveld et al., 2012; Bosveld et al., 2016; Merkel, 2014; Merkel et al., 2014; Rezakhaniha et al., 2012; Sagner et al., 2012). Here we focus on two methods that require segmentation. The first method (Blankenship et al., 2006; Levayer, Pelissier-Monier, & Lecuit, 2011; Marcinkevicius, Fernandez-Gonzalez, & Zallen, 2009; Rauzi, Verant, Lecuit, & Lenne, 2008; Simoes Sde et al., 2010; Tamada, Farrell, & Zallen, 2012) measures the relative protein concentration at junctions as a function of their orientation. Practically, for each junction the average intensity and angle relative to a reference axis (e.g., anteroposterior, proximal-distal etc.) are measured. Junctions are assigned to different angular bins, and their intensities are averaged. The dominant axis (or axes) of polarity is revealed by plotting average intensities as a function of angle in a histogram. To compare data from different samples, the junctional intensities can be normalized by the average of all interface intensities within each image. To increase the dynamic range of the measured polarity, camera and cytoplasmic background can be subtracted before intensity measurements (Collinet et al., 2015; Heemskerk et al., 2014; Munjal, Philippe, Munro, & Lecuit, 2015; Simoes Sde et al., 2010; Tamada et al., 2012). Note that, although this method may appear local, because it uses cell−cell interfaces, the local information is lost in the binning process and the method rather provides the axial polarity of the whole tissue. This method has often been applied to measure polarity in the germband of *Drosophila* using angular bins between 0 and 90 degrees (Levayer et al., 2011; Rauzi et al., 2008; Simoes Sde et al., 2010; Tamada et al., 2012). We recommend, instead, binning between 0 and 180 to distinguish not only polarities with horizontal and vertical orientations (i.e., enrichment at interfaces with 0 or 90 degrees), but also those with diagonal orientations (i.e., enrichment on interfaces with 45 and 135 degrees) (Blankenship et al., 2006). A related, but segmentation-independent, method is provided by the tool "Orientation J." This uses an edge detection formula and displays the local edge orientation with an angle-dependent color (Matis, Russler-Germain, Hu, Tomlin, & Axelrod, 2014; Rezakhaniha et al., 2012).

As an alternative to an angular intensity plot, axial polarity can also be characterized using a mathematical object called "nematic" (see Mathematical Representations of Vectorial and Axial Polarity box). A nematic is composed of an orientation and a magnitude, much like a vector. The main difference is that the orientation of a nematic merely defines an axis, while the orientation of a vector defines an axis and "points into" one of the two possible directions (Figs. 1A,B). Throughout this chapter, we will denote nematics by bold symbols with a tilde: $\tilde{\boldsymbol{A}}$.

The axial polarity of a cell can be quantified by a nematic $\tilde{\boldsymbol{A}}$ similarly to the method illustrated above for vectorial polarity of a cell (Aigouy et al., 2010; Merkel et al., 2014; Sagner et al., 2012). The cell polarity $\tilde{\boldsymbol{A}}$ is computed as the sum of nematics defined for each pixel of the cell outline (Fig. 1P). The axis of the pixel nematic is obtained by connecting the pixel to the cell center (green line, pixel marked by magenta outline, Fig. 1P). The magnitude of the pixel nematic is a product between the pixel intensity and the angle interval between its two neighboring pixels (angle marked in red in Fig. 1P). Like for the vectorial method discussed above, the angular factor ensures that a cell with constant pixel intensity along its outline will have zero axial polarity, even if its shape is elongated. For comparison among different cells, the cellular polarity nematic $\tilde{\boldsymbol{A}}$ can be normalized by the sum of all pixel nematic magnitudes. Also in this case, the magnitude of the normalized cell nematic varies between zero and one.

Cell-based and interface-based methods typically yield in most cases similar results. However note that for strongly elongated cells, the results can differ.

MATHEMATICAL REPRESENTATIONS OF VECTORIAL AND AXIAL POLARITY

To represent polarity orientation and magnitude in 2-D, we recommend using vectors for vectorial polarity and nematics for axial polarity. A vector $\boldsymbol{p}$ can be represented by an arrow (Fig. 1A). It consists of two components $\boldsymbol{p} = (p_x, p_y)$, which correspond to its projections on x and y axes, respectively. Based on these components, the polarity direction ϕ and magnitude $|p|$ (i.e., vector length) can be computed as follows:

$$\phi = \arctan 2\left(p_y, p_x\right) \quad \text{and} \quad |p| = \sqrt{p_x^2 + p_y^2}.$$

Here, "arctan2" denotes the arcus tangens function with two arguments. Conversely, the vector components can be computed from direction and length using the formulas

$$p_x = |p|\cos(\phi) \quad \text{and} \quad p_y = |p|\sin(\phi).$$

It is easier to visualize vectors using the representation by angle ϕ and norm $|p|$, while the representation by components should be used for arithmetics. To average n vectors, one first adds them and then divides their sum by n. To add two vectors, $\boldsymbol{p}_1 = (p_{1,x}, p_{1,y})$ and $\boldsymbol{p}_2 = (p_{2,x}, p_{2,y})$, their respective components have to be added separately: $\boldsymbol{p}_1 + \boldsymbol{p}_2 = (p_{1,x} + p_{2,x}, p_{1,y} + p_{2,y})$. Similarly, to multiply (or divide) a vector by a number n, each of its components, p_x and p_y, has to be multiplied (or divided) by this number n.

A nematic $\tilde{A}$ can be represented by a bar (Fig. 1B) and is composed of two components, $\tilde{A}_{xx}$ and $\tilde{A}_{xy}$. Similar to a vector, these components can be computed from the polarity orientation angle ψ (i.e., the angle of the bar) and the polarity magnitude $|\tilde{A}|$ (the length of the bar):

$$\tilde{A}_{xx} = |\tilde{A}|\cos(2\psi) \quad \text{and} \quad \tilde{A}_{xy} = |\tilde{A}|\sin(2\psi),$$

and vice versa

$$\psi = \frac{1}{2}\arctan 2(\tilde{A}_{xy}, \tilde{A}_{xx}) \quad \text{and} \quad |\tilde{A}| = \sqrt{\tilde{A}_{xx}^2 + \tilde{A}_{xy}^2}.$$

The representation by angle and magnitude is used to visualize nematics, whereas the representation by components is used for arithmetics, which works as for vectors.

Note that a bar representing a nematic is the same when you rotate it by 180 degrees, whereas a vector has to be rotated by 360 degrees to be the same. Consequently, while vectors add up to zero whenever they have the same lengths and opposing directions, two nematics add up to zero whenever they have the same magnitude and are perpendicular to each other. As a related example, two vectors of the same length add up to a vector with double length whenever they have the same direction (+/−360 degrees), while for nematics this is true whenever their axes are the same (+/−180 degrees).

4.2 TISSUE DEFORMATION

4.2.1 Without segmentation

Tissue deformation can be computed from the flow field extracted by PIV (see Section 3.2). This method is fast and does not require segmentation, but does not provide information on the underlying cellular processes.

The underlying idea is that tissue deformation is induced by the relative motion of adjacent tissue regions, while a global translation does not affect tissue shape at all (Figs. 3B–D). For example, Fig. 3C shows an expansion of a tissue along the horizontal (x) axis. This deformation corresponds to a velocity field whose x component, v_x, increases towards the right. The corresponding rate of expansion can be measured by computing the relative increase of the velocity component v_x per length unit in x direction. Fig. 3D shows another example, where the velocity component v_x is constant along x direction, but increases in y direction. This corresponds to a so-called "simple shear deformation." The rate of simple shear is the relative increase of the velocity component v_x per length unit in y direction. In addition to these two kinds of deformations, also the y component of the velocity, v_y, may change in both x and y directions. The mathematical object that contains all four of the corresponding deformation rates is called "velocity gradient" and is used to describe local tissue deformation (see Velocity Gradient and Tissue Deformation box).

Deformations can be decomposed into a so-called "isotropic deformation component" and an "anisotropic deformation component" (Fig. 3E,F). This can be useful because certain cellular events, for instance T1 transitions, only contribute to one of the two, i.e., anisotropic deformation. Both components can be extracted from the velocity gradient (see Velocity Gradient and Tissue Deformation box for details). Isotropic deformation refers to tissue expansion (or shrinkage) in all directions with the same rate and is characterized by the isotropic deformation rate V.

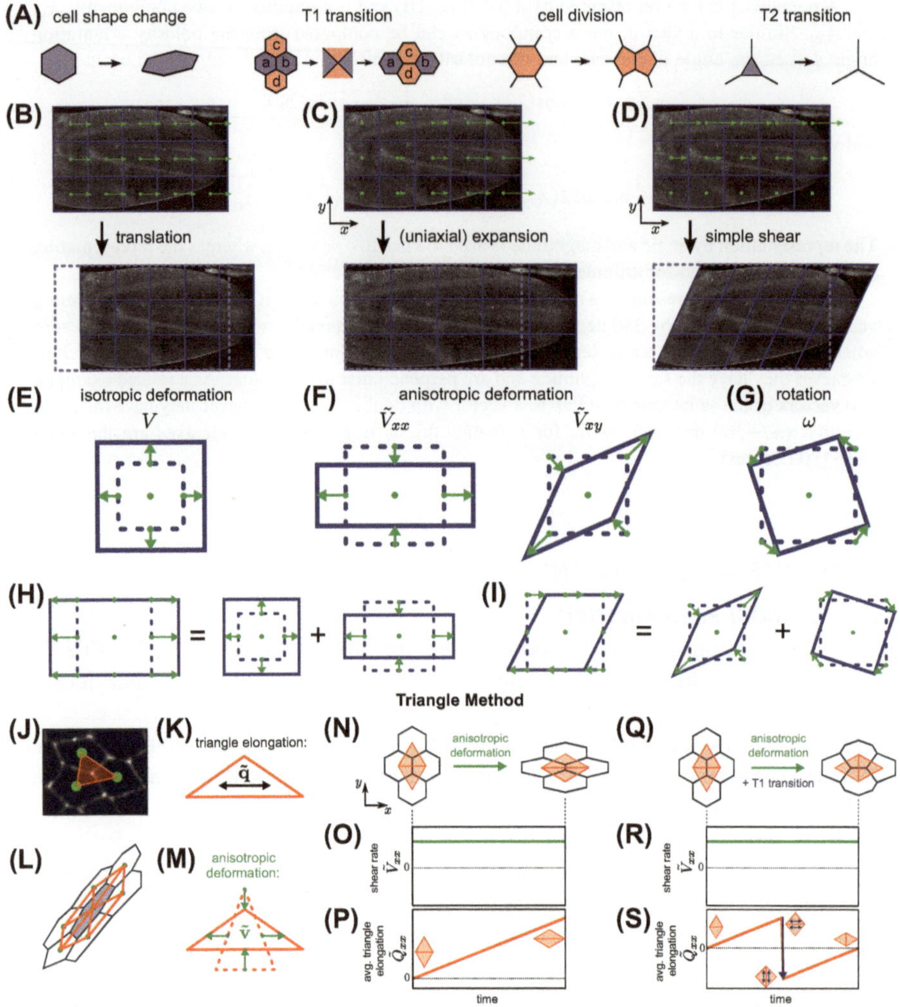

FIGURE 3 Quantification of large-scale tissue deformation and cell-scale contributions to it.

(A) Cellular processes underlying tissue morphogenesis. (B) A uniform translation does not deform the tissue. (C, D) Tissue deformation is created by spatial variation of the velocity. (C) A (uniaxial) expansion is created by increasing the horizontal component of the velocity along the *x* axis. (D) A simple shear deformation is created by increasing the horizontal component of the velocity along the *y* axis. (E–G) Based on the velocity gradient, the rates of (E) isotropic deformation, V, (F) anisotropic deformation, $\tilde{V}$ and (G) rotation, ω, can be computed. (H) Uniaxial expansion can be decomposed into an isotropic and an anisotropic part. (I) Simple shear deformation can be decomposed into an anisotropic deformation (i.e., pure shear) and a rotation. (J–S) Triangle Method to quantify the contributions of individual cellular events to tissue deformation. (J) A triangle (red) is created by connecting the respective centers (green) of three neighboring cells. (K) The elongation (i.e., shape anisotropy) of a single triangle is quantified by the nematic $\tilde{q}$. (L) The elongation of a cell

Anisotropic deformation, also called "pure shear," refers to a convergence-extension-like tissue deformation, where the area of the tissue remains constant (Fig. 3F). While isotropic deformation is described by the single parameter V, pure shear deformation is characterized by the rate and the axis of tissue expansion. It is thus characterized by a nematic $\tilde{V}$, like axial polarity (see Mathematical Representations of Vectorial and Axial Polarity box). In addition to isotropic and anisotropic deformation rates, the rotation rate ω of a piece of tissue can also be computed from the velocity gradient (Fig. 3G, see Velocity Gradient and Tissue Deformation box).

Note that any deformation can be described as a combination of the four deformation modes shown in Figs. 3E–G. For instance, expansion along the x axis is a combination of isotropic deformation and pure shear along the x axis (Fig. 3H). Moreover, simple shear is the combination of pure shear along a diagonal axis and rotation (Fig. 3I). Finally, beware that tissue deformations are typically spatially heterogeneous. The deformation rates V and $\tilde{V}$ and the rotation rate ω provide **local** information about tissue deformation.

VELOCITY GRADIENT AND TISSUE DEFORMATION

The velocity gradient is a combination of the respective gradients of both velocity components, v_x and v_y. For instance the gradient of v_x can be understood like a morphogen gradient: it has a direction and a magnitude (its steepness). Thus, it can be represented by a vector with two components $\partial v_x/\partial x$ and $\partial v_x/\partial y$, which are the respective derivatives (i.e., changes) of v_x in x and y direction. Hence, including v_y, the whole velocity gradient is composed of four different derivatives. These derivatives can be combined to compute the rates of isotropic deformation V, anisotropic deformation $\tilde{V}$ (which is a nematic, see Mathematical Representations of Vectorial and Axial Polarity box), and rotation ω:

$$V = \frac{\partial v_x}{\partial x} + \frac{\partial v_y}{\partial y}, \quad \tilde{V}_{xx} = \frac{1}{2}\left(\frac{\partial v_x}{\partial x} - \frac{\partial v_y}{\partial y}\right), \quad \tilde{V}_{xy} = \frac{1}{2}\left(\frac{\partial v_y}{\partial x} + \frac{\partial v_x}{\partial y}\right), \quad \omega = \frac{1}{2}\left(\frac{\partial v_y}{\partial x} - \frac{\partial v_x}{\partial y}\right).$$

With these definitions, the average rate of isotropic deformation V corresponds to the rate of relative area expansion (or shrinkage for negative V): $V = \dot{A_T}/A_T$, where A_T is the area of the tissue and the dot denotes the time derivative. Anisotropic deformation changes the aspect ratio of a tissue. In particular, if both anisotropic deformation and tissue elongation are oriented along the x axis, the horizontal component of the anisotropic deformation rate corresponds to the relative change of the tissue aspect ratio R : $\tilde{V}_{xx} = \dot{R}/(2R)$.

(blue) is computed by averaging the elongations $\tilde{q}$ of the triangles (red) that have one of their corners at the center of this cell. (M) The anisotropic deformation of a single triangle is quantified by the nematic $\tilde{v}$. (N–P) Pure shear of four cells along the horizontal axis with a constant pure shear rate (O) leads to a linear increase in the average triangle elongation (P). (Q–S) Pure shear deformation with a constant rate (R) during which a T1 transition occurs. Because the net cell elongation does not change between beginning and end (Q), the T1 transition accounts for the whole pure shear deformation. Note that the cell neighbor exchange removes and adds two triangles (Q,S). This induces an instantaneous reduction of the average triangle elongation along the horizontal axis at the instant of the neighbor exchange (schemes and *blue arrow* in S). Note that panels O, P, R, and S show the horizontal component of pure shear rate and average triangle elongation.

4.2.2 With segmentation

PIV can be used to characterize tissue deformations, but not the underlying cellular events. To measure the contributions of individual cellular processes to tissue deformation, time-lapse movies need to be segmented and cells tracked over time (see Section 3.4).

The contribution of individual cellular processes to both isotropic and anisotropic tissue deformation can be precisely quantified. Isotropic tissue deformation can be due to cell area changes, cell divisions, and cell extrusions. More precisely, the rate of isotropic tissue deformation can be decomposed into the relative change of average cell area $\overline{a}$ plus the normalized rate of cell divisions k_d (number of divisions per cell per time unit) minus the normalized rate of cell extrusions k_e (number of extrusions per cell per time unit) (Etournay et al., 2015; Merkel, 2014; Merkel et al., 2016):

$$V = \frac{1}{\overline{a}}\dot{\overline{a}} + k_d - k_e. \tag{1}$$

This remains true also when divisions occur without cell area growth.

Decomposing the anisotropic part of tissue deformation into cellular contributions is more complex and several methods have been used (Blanchard et al., 2009; Brodland et al., 2006; Butler et al., 2009; Economou, Brock, Cobourne, & Green, 2013; Etournay et al., 2015; Graner et al., 2008; Kabla, Blanchard, Adams, & Mahadevan, 2010; Merkel, 2014; Merkel et al., 2016; Rauzi et al., 2008). For instance, Blanchard et al. (2009) and Kabla et al. (2010) quantify cell elongation by fitting ellipses to cell shapes. Note that throughout this section, "elongation" refers to shape anisotropy, not the process of elongating. Tissue deformation is obtained for small patches of cells by linearly fitting the movement of all cell centers within a given patch. The contribution of cell intercalation is quantified for each patch by subtracting the change of average cell elongation from the total patch deformation.

Another method by Graner et al. (Graner et al., 2008; Guirao et al., 2015) describes the elongation of a cell using the connection lines between its center and the centers of all neighboring cells. Tissue deformation induces continuous changes in these connection lines. In contrast, T1 transitions, cell divisions, and cell extrusions induce rewiring of the connections. By comparing the connection lines before and after the respective topological transition, the contribution of the transition to tissue deformation can be extracted.

In the following, we focus on the so-called "Triangle Method" (Etournay et al., 2015; Merkel, 2014; Merkel et al., 2016). First, the centers of neighboring cells are connected by triangles (Fig. 3J). The elongation of each triangle is quantified by a nematic $\tilde{\boldsymbol{q}}$ (Fig. 3K), which can be used as proxy to define cell elongation (Fig. 3L). Upon cell motion and deformation, each triangle deforms with an anisotropic deformation rate $\tilde{\nu}$ (i.e., pure shear rate, Fig. 3M), which corresponds to the change of triangle elongation $\tilde{\boldsymbol{q}}$. The average triangle pure shear rate corresponds to the overall anisotropic tissue deformation rate $\tilde{\boldsymbol{V}}$.

The triangles are used to define the respective contributions of the different cellular events to the tissue pure shear rate $\tilde{\boldsymbol{V}}$ (Fig. 3A). In the simple case where all cells deform in the same way without topological transitions, the overall tissue pure shear rate $\tilde{V}$ corresponds to the change of the average triangle elongation, which we denote by $\tilde{Q} : \tilde{V} = \mathrm{D}\tilde{Q}/\mathrm{D}t$. An example is given in Fig. 3N, where four cells elongate horizontally with a constant shear rate $\tilde{V}$ (Fig. 3O), reflected in the linear increase of the average triangle elongation $\tilde{Q}$ (Fig. 3P).

Fig. 3Q illustrates how the contribution of T1 transitions to the pure shear rate is quantified by the Triangle Method. Shown is a process with a constant shear rate along the horizontal axis with a T1 transition. As there is no net change in the cell elongation (i.e., the average triangle elongation) (Fig. 3Q), the T1 transition has to fully account for the deformation during this process. In particular, at the instant of the cell neighbor exchange, two triangles disappear and two new triangles appear (Figs. 3Q,S). Consequently, the average triangle elongation along the horizontal axis suddenly decreases (schemes and blue arrow in Fig. 3S). This decrease in triangle elongation quantifies the overall contribution of the T1 transition to pure shear. Note that also rosette formation and resolution can be accounted for in this way (details on how to handle vertices with more than three cells are provided in Etournay et al., 2015; Merkel, 2014; Merkel et al., 2016).

To define contributions by cell divisions and T2 transitions to anisotropic deformation, the Triangle Method follows the same principle. In particular, during a cell division two triangles appear, whereas during a T2 transition, two triangles disappear.

Finally, the Triangle Method reveals an additional contribution to anisotropic tissue deformation, which is due to collective cell motion. In general, this contribution appears whenever triangle elongation and local tissue rotation or local area expansion vary across a tissue (see details in Etournay et al., 2015; Merkel, 2014; Merkel et al., 2016).

All three methods described here require segmentation and cell tracking. While the method by Blanchard et al. relies on cell outlines, the other two methods use only information about cell centers and cell neighborship. The first method represents cell shape more directly, but does not take into account cell divisions and cell extrusions, while the other two methods do. Finally, the Triangle Method is so far the only method that captures the contribution by collective cell motion.

A software suite that provides an implementation of the Triangle Method is available as part of the free tool "TissueMiner." This tool together with hands-on explanations and step-by-step tutorials can be found in (Etournay et al., 2016).

Finally, note that also Heemskerk et al. (2014) use triangles to quantify tissue deformation. However there the triangles are much larger; triangle corners are defined by the centers of clones. This method is thus suited to quantify large-scale tissue deformation. In contrast, since the Triangle Method defines the triangles at the cellular scale, the deformation induced by each individual cellular event can be precisely measured.

4.3 AVERAGING OF TISSUE DEFORMATION AND POLARITY PATTERNS

Local averaging of deformation and cell- or clone-based polarity can be used to reveal their large-scale patterns and improve S/N ratio (Aigouy et al., 2010; Guirao et al., 2015; Merkel et al., 2014; Sagner et al., 2012). This can be done by dividing the image into blocks and by averaging measured quantities within each block. Note that to average oriented objects such as vectors or nematics, their components need to be averaged separately. In particular, angles should never be averaged. Such blocks should be sufficiently large to ensure a good S/N ratio, but also small enough to preserve local differences between neighboring regions of the tissue. Note that averaging over several samples requires their alignment in space and time using tissue-specific landmarks (e.g., morphological landmarks, characteristic cell and tissue behaviors, etc. Aigouy et al., 2010; Collinet et al., 2015; Guirao et al., 2015; Merkel et al., 2014; Sagner et al., 2012).

CONCLUDING REMARKS

Cell deformations, divisions, rearrangements, and extrusions are well-known to participate in epithelial morphogenesis. In many cases, these events are coordinated within an epithelium through the action of planar polarized proteins. However, the extent to which each of these events contributes to overall tissue reshaping is not always obvious. In particular, the fact that a cellular event is frequent does not necessarily imply that it significantly contributes to morphogenesis. This highlights the importance of the development of new approaches to quantitatively study epithelial tissue deformations and planar polarity from the cell to tissue scale. In this chapter, we outline biological tools, imaging approaches, and computational methods to quantify the individual contributions of cellular events to morphogenesis. These enhance our understanding of epithelial morphogenesis by allowing direct summation over all the individual events that occur at the cellular level, and even linking them to molecular mechanisms. Finally, such precise quantifications allow to refine and test quantitative models that are expected to broaden our understanding of this fundamental biological process.

ACKNOWLEDGMENTS

We thank members of the Lecuit lab, for discussions. C.C. was supported by a Human Frontiers Science Program Long-Term Fellowship and by a postdoctoral fellowship of the Fondation Recherche Medicale. M.M. was supported by the Alfred P. Sloan Foundation, the Gordon and Betty Moore Foundation, and NSF-DMR-1352184. A.S. was supported by a Human Frontiers Science Program Long-Term Fellowship, Marie Curie Intra-European Fellowship, and EMBO Long-Term Fellowship.

REFERENCES

Aigouy, B., Farhadifar, R., Staple, D. B., Sagner, A., Roper, J. C., Julicher, F., & Eaton, S. (2010). Cell flow reorients the axis of planar polarity in the wing epithelium of *Drosophila*. *Cell, 142*(5), 773–786. http://dx.doi.org/10.1016/j.cell.2010.07.042.

Ambegaonkar, A. A., & Irvine, K. D. (2015). Coordination of planar cell polarity pathways through Spiny legs. *eLife, 4*, e09946. http://dx.doi.org/10.7554/eLife.09946.

Baena-Lopez, L. A., Baonza, A., & Garcia-Bellido, A. (2005). The orientation of cell divisions determines the shape of *Drosophila* organs. *Current Biology: CB, 15*(18), 1640–1644. http://dx.doi.org/10.1016/j.cub.2005.07.062.

Bardet, P., Guirao, B., Paoletti, C., Serman, F., Léopold, V., Bosveld, F., … Bellaïche, Y. (2013). PTEN controls junction lengthening and stability during cell rearrangement in epithelial tissue. *Developmental Cell, 25*(5), 534–546. http://dx.doi.org/10.1016/j.devcel.2013.04.020.

Bertet, C., Sulak, L., & Lecuit, T. (2004). Myosin-dependent junction remodelling controls planar cell intercalation and axis elongation. *Nature, 429*(6992), 667–671. http://dx.doi.org/10.1038/nature02590.

Besson, C., Bernard, F., Corson, F., Rouault, H., Reynaud, E., Keder, A., … Schweisguth, F. (2015). Planar cell polarity breaks the symmetry of PAR protein distribution prior to mitosis in *Drosophila* sensory organ precursor cells. *Current Biology: CB, 25*(8), 1104–1110. http://dx.doi.org/10.1016/j.cub.2015.02.073.

Blanchard, G. B., Kabla, A. J., Schultz, N. L., Butler, L. C., Sanson, B., Gorfinkiel, N., … Adams, R. J. (2009). Tissue tectonics: Morphogenetic strain rates, cell shape change and intercalation. *Nature Methods, 6*, 458–464. http://dx.doi.org/10.1038/nmeth.1327.

Blankenship, J. T., Backovic, S. T., Sanny, J. S., Weitz, O., & Zallen, J. A. (2006). Multicellular rosette formation links planar cell polarity to tissue morphogenesis. *Developmental Cell, 11*(4), 459–470. http://dx.doi.org/10.1016/j.devcel.2006.09.007.

Bohn, S. (2003). Bubbles under stress. *European Physical Journal E Soft Matter, 11*(2), 177–189. http://dx.doi.org/10.1140/epje/i2003-10014-x.

Bosveld, F., Bonnet, I., Guirao, B., Tlili, S., Wang, Z., Petitalot, A., … Bellaiche, Y. (2012). Mechanical control of morphogenesis by fat/dachsous/four-jointed planar cell polarity pathway. *Science, 336*, 724–727. http://dx.doi.org/10.1126/science.1221071.

Bosveld, F., Markova, O., Guirao, B., Martin, C., Wang, Z., Pierre, A., … Bellaïche, Y. (2016). Epithelial tricellular junctions act as interphase cell shape sensors to orient mitosis. *Nature, 530*(7591), 495–498. http://dx.doi.org/10.1038/nature16970.

Brittle, A., Thomas, C., & Strutt, D. (2012). Planar polarity specification through asymmetric subcellular localization of fat and dachsous. *Current Biology: CB, 22*, 907–914. http://dx.doi.org/10.1016/j.cub.2012.03.053.

Brodland, G. W., Chen, D. I. L., & Veldhuis, J. H. (2006). A cell-based constitutive model for embryonic epithelia and other planar aggregates of biological cells. *International Journal of Plasticity, 22*, 965–995. http://dx.doi.org/10.1016/j.ijplas.2005.05.002.

Butler, L. C., Blanchard, G. B., Kabla, A. J., Lawrence, N. J., Welchman, D. P., Mahadevan, L., … Sanson, B. (2009). Cell shape changes indicate a role for extrinsic tensile forces in *Drosophila* germ-band extension. *Nature Cell Biology, 11*, 859–864. http://dx.doi.org/10.1038/ncb1894.

Campinho, P., Behrndt, M., Ranft, J., Risler, T., Minc, N., & Heisenberg, C. P. (2013). Tension-oriented cell divisions limit anisotropic tissue tension in epithelial spreading during zebrafish epiboly. *Nature Cell Biology, 15*(12), 1405–1414. http://dx.doi.org/10.1038/ncb2869.

Chacon-Heszele, M. F., Ren, D., Reynolds, A. B., Chi, F., & Chen, P. (2012). Regulation of cochlear convergent extension by the vertebrate planar cell polarity pathway is dependent on p120-catenin. *Development, 139*(5), 968–978. http://dx.doi.org/10.1242/dev.065326.

Chae, J., Kim, M. J., Goo, J. H., Collier, S., Gubb, D., Charlton, J., … Park, W. J. (1999). The *Drosophila* tissue polarity gene starry night encodes a member of the protocadherin family. *Development, 126*(23), 5421–5429.

Cilla, R., Mechery, V., Hernandez de Madrid, B., Del Signore, S., Dotu, I., & Hatini, V. (2015). Segmentation and tracking of adherens junctions in 3D for the analysis of epithelial tissue morphogenesis. *PLoS Computational Biology, 11*(4), e1004124. http://dx.doi.org/10.1371/journal.pcbi.1004124.

Collinet, C., Rauzi, M., Lenne, P. F., & Lecuit, T. (2015). Local and tissue-scale forces drive oriented junction growth during tissue extension. *Nature Cell Biology, 17*(10), 1247–1258. http://dx.doi.org/10.1038/ncb3226.

Conchello, J., & Lichtman, J. W. (2005). Optical sectioning microscopy. *Nature Methods, 2*, 920–931.

Danielian, P. S., Muccino, D., Rowitch, D. H., Michael, S. K., & McMahon, A. P. (1998). Modification of gene activity in mouse embryos in utero by a tamoxifen-inducible form of Cre recombinase. *Current Biology: CB, 8*(24), 1323–1326.

Denk, W., Strickler, J. H., & Webb, W. W. (1990). Two-photon laser scanning fluorescence microscopy. *Science, 248*(4951), 73–76.

Denk, W., & Svoboda, K. (1997). Photon upmanship: Why multiphoton imaging is more than a gimmick. *Neuron, 18*(3), 351–357.

Devenport, D., Oristian, D., Heller, E., & Fuchs, E. (2011). Mitotic internalization of planar cell polarity proteins preserves tissue polarity. *Nature Cell Biology, 13*, 893–902. http://dx.doi.org/10.1038/ncb2284.

Economou, A. D., Brock, L. J., Cobourne, M. T., & Green, J. B.a (2013). Whole population cell analysis of a landmark-rich mammalian epithelium reveals multiple elongation mechanisms. *Development, 140*, 4740–4750. http://dx.doi.org/10.1242/dev.096545.

Eisenhoffer, G. T., Loftus, P. D., Yoshigi, M., Otsuna, H., Chien, C. B., Morcos, P. A., & Rosenblatt, J. (2012). Crowding induces live cell extrusion to maintain homeostatic cell numbers in epithelia. *Nature, 484*(7395), 546–549. http://dx.doi.org/10.1038/nature10999.

Eisenhoffer, G. T., & Rosenblatt, J. (2013). Bringing balance by force: Live cell extrusion controls epithelial cell numbers. *Trends Cell Biology, 23*(4), 185–192. http://dx.doi.org/10.1016/j.tcb.2012.11.006.

Etournay, R., Merkel, M., Popović, M., Brandl, H., Dye, N. A., Aigouy, B., ... Jülicher, F. (2016). TissueMiner: A multiscale analysis toolkit to quantify how cellular processes create tissue dynamics. *eLife, 5*. http://dx.doi.org/10.7554/eLife.14334.

Etournay, R., Popović, M., Merkel, M., Nandi, A., Blasse, C., Aigouy, B., ... Eaton, S. (2015). Interplay of cell dynamics and epithelial tension during morphogenesis of the *Drosophila* pupal wing. *eLife, 4*, e07090. http://dx.doi.org/10.7554/eLife.07090.

Forster, B., Van De Ville, D., Berent, J., Sage, D., & Unser, M. (2004). Complex wavelets for extended depth-of-field: A new method for the fusion of multichannel microscopy images. *Microscopy Research and Technique, 65*(1–2), 33–42. http://dx.doi.org/10.1002/jemt.20092.

Ganguly, S., Williams, L. S., Palacios, I. M., & Goldstein, R. E. (2012). Cytoplasmic streaming in *Drosophila* oocytes varies with kinesin activity and correlates with the microtubule cytoskeleton architecture. *Proceedings of the National Academy of Sciences of the United States of America, 109*(38), 15109–15114. http://dx.doi.org/10.1073/pnas.1203575109.

Gelbart, M. A., He, B., Martin, A. C., Thiberge, S. Y., Wieschaus, E. F., & Kaschube, M. (2012). Volume conservation principle involved in cell lengthening and nucleus movement during tissue morphogenesis. *Proceedings of the National Academy of Sciences of the United States of America, 109*(47), 19298–19303. http://dx.doi.org/10.1073/pnas.1205258109.

Golic, K. G. (1991). Site-specific recombination between homologous chromosomes in *Drosophila*. *Science, 252*(5008), 958–961.

Goodrich, L. V., & Strutt, D. (2011). Principles of planar polarity in animal development. *Development, 138*(10), 1877–1892. http://dx.doi.org/10.1242/dev.054080.

Graner, F., Dollet, B., Raufaste, C., & Marmottant, P. (2008). Discrete rearranging disordered patterns, part I: Robust statistical tools in two or three dimensions. *The European Physical Journal E, 25*, 349–369. http://dx.doi.org/10.1140/epje/i2007-10298-8.

Guillot, C., & Lecuit, T. (2013). Mechanics of epithelial tissue homeostasis and morphogenesis. *Science, 340*(6137), 1185–1189. http://dx.doi.org/10.1126/science.1235249.

Guirao, B., Rigaud, S. U., Bosveld, F., Bailles, A., López-Gay, J., Ishihara, S., … Bellaïche, Y. (2015). Unified quantitative characterization of epithelial tissue development. *eLife, 4*. http://dx.doi.org/10.7554/eLife.08519.

Hayashi, S., & McMahon, A. P. (2002). Efficient recombination in diverse tissues by a tamoxifen-inducible form of Cre: A tool for temporally regulated gene activation/inactivation in the mouse. *Developmental Biology, 244*(2), 305–318. http://dx.doi.org/10.1006/dbio.2002.0597.

He, B., Doubrovinski, K., Polyakov, O., & Wieschaus, E. (2014). Apical constriction drives tissue-scale hydrodynamic flow to mediate cell elongation. *Nature, 508*(7496), 392–396. http://dx.doi.org/10.1038/nature13070.

Heemskerk, I., Lecuit, T., & LeGoff, L. (2014). Dynamic clonal analysis based on chronic in vivo imaging allows multiscale quantification of growth in the *Drosophila* wing disc. *Development, 141*(11), 2339–2348. http://dx.doi.org/10.1242/dev.109264.

Heller, D., Hoppe, A., Restrepo, S., Gatti, L., Tournier, A. L., Tapon, N., … Mao, Y. (2016). EpiTools: An open-source image analysis toolkit for quantifying epithelial growth dynamics. *Developmental Cell, 36*(1), 103–116. http://dx.doi.org/10.1016/j.devcel.2015.12.012.

Helmchen, F., & Denk, W. (2005). Deep tissue two-photon microscopy. *Nature Methods, 2*(12), 932–940. http://dx.doi.org/10.1038/nmeth818.

Huisken, J., & Stainier, D. Y. (2007). Even fluorescence excitation by multidirectional selective plane illumination microscopy (mSPIM). *Optics Letters, 32*(17), 2608–2610.

Huisken, J., & Stainier, D. Y. (2009). Selective plane illumination microscopy techniques in developmental biology. *Development, 136*(12), 1963–1975. http://dx.doi.org/10.1242/dev.022426.

Inoué, S., & Inoué, T. (2002). Direct-view high-speed confocal scanner: the CSU-10. *Methods in Cell Biology, 70*, 87–127.

Irvine, K. D., & Wieschaus, E. (1994). Cell intercalation during *Drosophila* germband extension and its regulation by pair-rule segmentation genes. *Development, 120*(4), 827–841.

Kabla, A., Blanchard, G., Adams, R., & Mahadevan, L. (2010). Bridging cell and tissue behavior in embryo development. In *Cell mechanics: From single scale-based models to multiscale modeling*. Chapman & Hall.

Karner, C. M., Chirumamilla, R., Aoki, S., Igarashi, P., Wallingford, J. B., & Carroll, T. J. (2009). Wnt9b signaling regulates planar cell polarity and kidney tubule morphogenesis. *Nature Genetics, 41*(7), 793–799. http://dx.doi.org/10.1038/ng.400.

Keller, P. J., & Stelzer, E. H. (2008). Quantitative in vivo imaging of entire embryos with digital scanned laser light sheet fluorescence microscopy. *Current Opinion in Neurobiology, 18*(6), 624–632. http://dx.doi.org/10.1016/j.conb.2009.03.008.

Khan, Z., Wang, Y. C., Wieschaus, E. F., & Kaschube, M. (2014). Quantitative 4D analyses of epithelial folding during *Drosophila* gastrulation. *Development, 141*(14), 2895–2900. http://dx.doi.org/10.1242/dev.107730.

Kretzschmar, K., & Watt, F. M. (2012). Lineage tracing. *Cell, 148*(1–2), 33–45. http://dx.doi.org/10.1016/j.cell.2012.01.002.

Krzic, U., Gunther, S., Saunders, T. E., Streichan, S. J., & Hufnagel, L. (2012). Multiview light-sheet microscope for rapid in toto imaging. *Nature Methods, 9*(7), 730–733. http://dx.doi.org/10.1038/nmeth.2064.

Legoff, L., Rouault, H., & Lecuit, T. (2013). A global pattern of mechanical stress polarizes cell divisions and cell shape in the growing *Drosophila* wing disc. *Development, 140*(19), 4051–4059. http://dx.doi.org/10.1242/dev.090878.

Leung, C. Y., & Fernandez-Gonzalez, R. (2015). Quantitative image analysis of cell behavior and molecular dynamics during tissue morphogenesis. *Methods in Molecular Biology, 1189*, 99–113. http://dx.doi.org/10.1007/978-1-4939-1164-6_7.

Levayer, R., & Lecuit, T. (2013). Oscillation and polarity of E-cadherin asymmetries control actomyosin flow patterns during morphogenesis. *Developmental Cell, 26*(2), 162–175. http://dx.doi.org/10.1016/j.devcel.2013.06.020.

Levayer, R., Pelissier-Monier, A., & Lecuit, T. (2011). Spatial regulation of Dia and Myosin-II by RhoGEF2 controls initiation of E-cadherin endocytosis during epithelial morphogenesis. *Nature Cell Biology, 13*(5), 529–540. http://dx.doi.org/10.1038/ncb2224.

Lienkamp, S. S., Liu, K., Karner, C. M., Carroll, T. J., Ronneberger, O., Wallingford, J. B., & Walz, G. (2012). Vertebrate kidney tubules elongate using a planar cell polarity-dependent, rosette-based mechanism of convergent extension. *Nature Genetics, 44*(12), 1382–1387. http://dx.doi.org/10.1038/ng.2452.

Lynch, H. E., Veldhuis, J., Brodland, G. W., & Hutson, M. S. (2014). Modeling cell elongation during germ band retraction: Cell autonomy versus applied anisotropic stress. *New Journal of Physics, 16*(2014), 055003. http://dx.doi.org/10.1088/1367-2630/16/5/055003.

Marcinkevicius, E., Fernandez-Gonzalez, R., & Zallen, J. A. (2009). Q&A: Quantitative approaches to planar polarity and tissue organization. *Journal of Biology, 8*(12), 103. http://dx.doi.org/10.1186/jbiol191.

Marinari, E., Mehonic, A., Curran, S., Gale, J., Duke, T., & Baum, B. (2012). Live-cell delamination counterbalances epithelial growth to limit tissue overcrowding. *Nature, 484*(7395), 542–545. http://dx.doi.org/10.1038/nature10984.

Matis, M., Axelrod, J. D., & Galic, M. (2012). A Universal analysis tool for the detection of asymmetric signal distribution in microscopic images. *Developmental Dynamics*, 1301–1309. http://dx.doi.org/10.1002/dvdy.23818.

Matis, M., Russler-Germain, D.a., Hu, Q., Tomlin, C. J., & Axelrod, J. D. (2014). Microtubules provide directional information for core PCP function. *eLife, 3*, e02893. http://dx.doi.org/10.7554/eLife.02893.

Mayer, M., Depken, M., Bois, J. S., Julicher, F., & Grill, S. W. (2010). Anisotropies in cortical tension reveal the physical basis of polarizing cortical flows. *Nature, 467*(7315), 617–621. http://dx.doi.org/10.1038/nature09376.

de Medeiros, G., Norlin, N., Gunther, S., Albert, M., Panavaite, L., Fiuza, U. M., … Hufnagel, L. (2015). Confocal multiview light-sheet microscopy. *Nature Communications, 6*, 8881. http://dx.doi.org/10.1038/ncomms9881.

Merkel, M. (2014). *From cells to tissues: Remodeling and polarity reorientation in epithelial tissues.*

Merkel, M., Etounay, R., Popović, M., Salbreux, G., Eaton, S., & Jülicher, F. (2016). Triangles bridge the scales: Quantifying cellular contributions to tissue deformation. arXiv: 1607.00357. https://arxiv.org/abs/1607.00357.

Merkel, M., Sagner, A., Gruber, F. S., Etournay, R., Blasse, C., Myers, E., … Julicher, F. (2014). The balance of prickle/spiny-legs isoforms controls the amount of coupling between core and fat PCP systems. *Current Biology: CB, 24*(18), 2111–2123. http://dx.doi.org/10.1016/j.cub.2014.08.005.

Mosaliganti, K. R., Noche, R. R., Xiong, F., Swinburne, I. A., & Megason, S. G. (2012). ACME: Automated cell morphology extractor for comprehensive reconstruction of cell membranes. *PLoS Computational Biology, 8*(12), e1002780. http://dx.doi.org/10.1371/journal.pcbi.1002780.

Munjal, A., Philippe, J. M., Munro, E., & Lecuit, T. (2015). A self-organized biomechanical network drives shape changes during tissue morphogenesis. *Nature, 524*(7565), 351–355. http://dx.doi.org/10.1038/nature14603.

Nagy, A. (2000). Cre recombinase: The universal reagent for genome tailoring. *Genesis, 26*(2), 99–109.

Nishimura, T., Honda, H., & Takeichi, M. (2012). Planar cell polarity links axes of spatial dynamics in neural-tube closure. *Cell, 149*(5), 1084–1097. http://dx.doi.org/10.1016/j.cell.2012.04.021.

Nishimura, T., & Takeichi, M. (2008). Shroom3-mediated recruitment of Rho kinases to the apical cell junctions regulates epithelial and neuroepithelial planar remodeling. *Development, 135*(8), 1493–1502. http://dx.doi.org/10.1242/dev.019646.

Pal, C., Chakrabarti, A., & Ghosh, R. (2015). *A Brief survey of recent edge-preserving smoothing algorithms on digital images*. Procedia Computer Science, ArXiv e-prints.

Preibisch, S., Saalfeld, S., & Tomancak, P. (2009). Globally optimal stitching of tiled 3D microscopic image acquisitions. *Bioinformatics, 25*(11), 1463–1465. http://dx.doi.org/10.1093/bioinformatics/btp184.

Raffel, M. (2007). *Particle image velocimetry: A practical guide* (2nd ed.). Heidelberg; New York: Springer.

Rauzi, M., Verant, P., Lecuit, T., & Lenne, P.-F. (2008). Nature and anisotropy of cortical forces orienting *Drosophila* tissue morphogenesis. *Nature Cell Biology, 10*, 1401–1410. http://dx.doi.org/10.1038/ncb1798.

Rezakhaniha, R., Agianniotis, A., Schrauwen, J. T. C., Griffa, A., Sage, D., Bouten, C. V. C., … Stergiopulos, N. (2012). Experimental investigation of collagen waviness and orientation in the arterial adventitia using confocal laser scanning microscopy. *Biomechanics and Modeling in Mechanobiology, 11*, 461–473. http://dx.doi.org/10.1007/s10237-011-0325-z.

Sagner, A., Merkel, M., Aigouy, B., Gaebel, J., Brankatschk, M., Jülicher, F., & Eaton, S. (2012). Establishment of global patterns of planar polarity during growth of the *Drosophila* wing epithelium. *Current Biology: CB, 22*, 1296–1301. http://dx.doi.org/10.1016/j.cub.2012.04.066.

Sauer, B., & Henderson, N. (1988). Site-specific DNA recombination in mammalian cells by the Cre recombinase of bacteriophage P1. *Proceedings of the National Academy of Sciences of the United States of America, 85*(14), 5166–5170.

Schindelin, J., Arganda-Carreras, I., Frise, E., Kaynig, V., Longair, M., Pietzsch, T., … Cardona, A. (2012). Fiji: An open-source platform for biological-image analysis. *Nature Methods, 9*(7), 676–682. http://dx.doi.org/10.1038/nmeth.2019.

Schneider, C. A., Rasband, W. S., & Eliceiri, K. W. (2012). NIH Image to ImageJ: 25 years of image analysis. *Nature Methods, 9*(7), 671–675.

Simoes Sde, M., Blankenship, J. T., Weitz, O., Farrell, D. L., Tamada, M., Fernandez-Gonzalez, R., & Zallen, J. A. (2010). Rho-kinase directs Bazooka/Par-3 planar polarity during *Drosophila* axis elongation. *Developmental Cell, 19*(3), 377–388. http://dx.doi.org/10.1016/j.devcel.2010.08.011.

So, P. T., Dong, C. Y., Masters, B. R., & Berland, K. M. (2000). Two-photon excitation fluorescence microscopy. *Annual Review of Biomedical Engineering, 2*, 399–429. http://dx.doi.org/10.1146/annurev.bioeng.2.1.399.

Srinivas, S., Watanabe, T., Lin, C. S., William, C. M., Tanabe, Y., Jessell, T. M., & Costantini, F. (2001). Cre reporter strains produced by targeted insertion of EYFP and ECFP into the ROSA26 locus. *BMC Developmental Biology, 1*, 4.

Stojmenovic, M., & Nayak, A. (2007). Direct ellipse fitting and measuring based on shape boundaries. *Advances in Image and Video Technology, Proceedings, 4872*, 221–235.

Supatto, W., Debarre, D., Moulia, B., Brouzes, E., Martin, J. L., Farge, E., & Beaurepaire, E. (2005). In vivo modulation of morphogenetic movements in *Drosophila* embryos with femtosecond laser pulses. *Proceedings of the National Academy of Sciences of the United States of America, 102*(4), 1047–1052. http://dx.doi.org/10.1073/pnas.0405316102.

Tada, M., & Kai, M. (2012). Planar cell polarity in coordinated and directed movements. *Current Topics in Developmental Biology, 101*, 77–110. http://dx.doi.org/10.1016/B978-0-12-394592-1.00004-1.

Tamada, M., Farrell, D. L., & Zallen, J. A. (2012). Abl regulates planar polarized junctional dynamics through beta-catenin tyrosine phosphorylation. *Developmental Cell, 22*(2), 309–319. http://dx.doi.org/10.1016/j.devcel.2011.12.025.

Tanaami, T., Otsuki, S., Tomosada, N., Kosugi, Y., Shimizu, M., & Ishida, H. (2002). High-speed 1-frame/ms scanning confocal microscope with a microlens and Nipkow disks. *Applied Optics, 41*, 4704–4708.

Usui, T., Shima, Y., Shimada, Y., Hirano, S., Burgess, R. W., Schwarz, T. L., … Uemura, T. (1999). Flamingo, a seven-pass transmembrane cadherin, regulates planar cell polarity under the control of Frizzled. *Cell, 98*(5), 585–595. S0092-8674(00)80046-X [pii].

Vincent, L., & Soille, P. (1991). Watersheds in digital spaces - an efficient algorithm based on immersion simulations. *IEEE Transactions on Pattern Analysis and Machine Intelligence, 13*(6), 583–598. http://dx.doi.org/10.1109/34.87344.

Walck-Shannon, E., & Hardin, J. (2014). Cell intercalation from top to bottom. *Nature Reviews. Molecular Cell Biology, 15*(1), 34–48. http://dx.doi.org/10.1038/nrm3723.

Wallingford, J. B. (2012). Planar cell polarity and the developmental control of cell behavior in vertebrate embryos. *Annual Review of Cell and Developmental Biology, 28*, 627–653. http://dx.doi.org/10.1146/annurev-cellbio-092910-154208.

Wang, E., Babbey, C. M., & Dunn, K. W. (2005). Performance comparison between the high-speed Yokogawa spinning disc confocal system and single-point scanning confocal systems. *Journal of Microscopy, 218*, 148–159.

Wang, J., Mark, S., Zhang, X., Qian, D., Yoo, S. J., Radde-Gallwitz, K., … Chen, P. (2005). Regulation of polarized extension and planar cell polarity in the cochlea by the vertebrate PCP pathway. *Nature Genetics, 37*(9), 980–985. http://dx.doi.org/10.1038/ng1622.

Weaire, D. L., & Hutzler, S. (2001). *The Physics of foams*.

Weber, M., & Huisken, J. (2011). Light sheet microscopy for real-time developmental biology. *Current Opinion in Genetics & Development, 21*(5), 566–572. http://dx.doi.org/10.1016/j.gde.2011.09.009.

Xu, T., & Rubin, G. M. (1993). Analysis of genetic mosaics in developing and adult *Drosophila* tissues. *Development, 117*(4), 1223–1237.

Zallen, J. A., & Wieschaus, E. (2004). Patterned gene expression directs bipolar planar polarity in *Drosophila*. *Developmental Cell, 6*(3), 343–355.

CHAPTER

Probing tissue interaction with laser-based cauterization in the early developing *Drosophila* embryo

8

M. Rauzi

Université Côte d'Azur, CNRS, INSERM, iBV, Nice, France

E-mail: matteo.rauzi@unice.fr

CHAPTER OUTLINE

Abstract

Tissue morphogenesis is governed by mechanical forces generated by cell cytoskeletal networks. It has been shown that subcellular forces are responsible for cell shape

Methods in Cell Biology, Volume 139, ISSN 0091-679X, http://dx.doi.org/10.1016/bs.mcb.2016.11.003

changes. Nevertheless cells in a developing organism do not act in isolation: cells contact and adhere one another, and forces are transmitted from cell-to-cell throughout tissues. Understanding how forces are integrated at the tissue level and finally at the full animal scale is nowadays a major challenge that will allow shedding new light on how embryo morphogenesis takes place. In this chapter, I present a new laser-based technique to probe tissue coupling in a living *Drosophila* embryo. Such technique allows generating mechanical fix boundaries that can eventually impair or modulate cell flows and tissue displacements to probe tissue interaction.

INTRODUCTION

The *Drosophila* embryo has been proven to be a great system to study tissue morphogenesis and mechanics since powerful genetic tools are available for this model organism. Nevertheless, the *Drosophila* embryo has some intrinsic properties that can be considered as major drawbacks for experimentalists. For instance the fly embryo is encapsulated in different compartments, one of which (the most inner one) is delimited by a membrane (named the vitelline membrane) that cannot be removed since it is necessary for the embryo survival. The vitelline membrane impairs a direct access to the cells of the embryo. As a consequence, powerful experimental techniques (e.g., cell transplantation and pipette-based force measurements) cannot be applied to the *Drosophila* embryo.

The vitelline membrane is a semipermeable layer that does not displace during the development of the embryo, and it allows to preserve a certain inner embryo pressure (Waring, 2000). Such a layer closely envelops the embryo surrounding the blastoderm epithelium at a distance of less than 0.5 μm. In this chapter, I present a new infrared laser–based technique that exploits the fact that the vitelline membrane is a static enveloping layer and that is in great proximity to the blastoderm epithelium. By exposing the apical region of the cells and the vitelline membrane to an infrared femtosecond laser, it is possible to generate local thermal effects that cauterize the tissue and fuse it to the vitelline membrane generating spatially defined fix boundaries. These cauterized regions act finally as barriers blocking or modulating cell flows and tissue displacements. This allows probing directly tissue mechanical interaction. This technique can then be coupled to a spinning disc imaging unit (Collinet, Rauzi, Lenne, & Lecuit, 2015) or to a multiview light sheet microscope (Rauzi et al., 2015) to allow a more local or in toto embryo visualization and analysis of the blastoderm epithelium, respectively.

1. MATERIALS

The materials listed here below are necessary to set up a system that allows performing laser cauterization on developing *Drosophila* embryos.

1.1 LASER, OPTICS, AND MECHANICAL COMPONENTS

1. A femtosecond Yb:YAG infrared laser with the following specification: 1 W average power, 1030 nm wavelength, <200 fs pulse, 50 MHz repetition rate (Mikan, Amplitudes Systemès, Pessac, France). Other lasers with similar specifications can also be used.
2. A mechanical shutter or a Pockel cell to turn on and off the cauterization light path.
3. A dampener block for class 4 lasers.
4. If not using a Pockel cell, build an attenuator composed by a Glan Laser and a polarization rotator λ/2, both optimized for 1030 nm wavelength and mounted on rotating supports: for the Glan Laser, the support can be rotated and fixed manually while for the λ/2 this should be electronically controlled.
5. A Pockel cell can be used in substitution to the shutter and the Glan Laser λ/2 attenuator.
6. Mirrors with coating for optimal infrared light reflection.
7. A 45-degree dichroic mirror reflecting the infrared light (1030 nm) and transmitting the visible light. The width of the dichroic mirror should be equal to the diameter ($1/e^2$) of the expended beam while the length should be roughly 1.5 times the width since this is used with a 45 degree angle.
8. Two or four plano-convex lenses with infrared coatings to engineer one or two telescopes to expand the laser beam in one or two steps accordingly.
9. For a spatially modulated highly flexible cauterization system, two galvanometric mirrors to control the x and y position of the beam on the specimen.
10. A power meter with definable wavelength and with a head detector adapted to high-power lasers (≥1 W).
11. An infrared visor and fluorescent paper discs to visualize the laser beam.
12. Optical table; optic holders; rails to reduce the number of degree of freedom of the optical components for an easier alignment; periscopes to elevate the beam when necessary; M6 and M4 screws.
13. High numerical aperture infrared correction objective (>1 NA) like for example the Nikon 25× 1.1 NA.

1.2 SOFTWARE

Use LabView software to control the electromechanical components via the controller of the shutter, of the λ/2 rotator and of the galvanometric mirrors.

1.3 SAMPLE PREPARATION

For laser alignment:

- If using a conventional spinning disc microscope: 0.17 mm slip covered with a film of fluorescent dye or with fluorescent beads with a diameter equal to or smaller than the resolution of the objective.

- If using a light sheet microscope with immersion chamber: PBS or fish medium (E3).

For biological experiments,

- For a spinning disc imaging system: 0.17 mm slip, bleach, heptane glue, and halocarbon oil (Cavey & Lecuit, 2008).
- For a light sheet microscope: capillary filled with agar or with other low scattering gels.

1.4 SAFETY MATERIAL

1. Goggles adapted for class 4 lasers (OD > 8). Optimally, goggles should be made of glass for long exposure protection.
2. Metal tubes to intubate the optic path.
3. Safety light indicators connected to the laser controller.

2. METHODS

Here are presented the methods to set up and apply the cauterization system. Such system can work in tandem with imaging scopes like a spinning disk microscope or a light sheet microscope: the former allows fast imaging with normal illumination for optimal uniform excitation while the latter allows fast imaging with tangential illumination for reduced photo bleaching. In this method, section is also briefly presented the EM imaging procedure to visualize and analyze at very high resolution the cauterized region after performing the experiment.

2.1 OPTIC PATH SETUP

Throughout this procedure, use the infrared visor and the fluorescent paper plate to visualize the laser beam. Remember to always wear protecting goggles while setting up the system.

1. Position and fix with clamps and screws the infrared femtosecond laser on the optic table. Position the shutter in front of the laser with the reflecting side of the shutting membrane facing the laser with a tilted angle.
2. Position and fix the laser dampener block so that the reflected light from the closed shutter is neutralized.
3. Position on the optical path and fix two coated mirrors sufficiently spaced one from the other. These should be mounted on tilting stages to provide sufficient degrees of freedom for laser alignment. The two mirrors should produce 90 degrees reflections of the laser beam.
4. Position and fix the Glan Laser in front of the reflected laser beam. This should be rotated until reaching maximum laser transmission. Use the power meter to

monitor laser power while rotating the Glan Laser. Fix the Glan Laser rotating holder.

5. Now, just before the Glan Laser, position the λ/2 plate. Make sure the laser beam travels roughly through the center of the λ/2 plate and of the Glan Laser. To align the beam use the degrees of freedom provided by the mirror's holders and two irises (**Note 1**).
6. Rotate the λ/2 plate to reduce the power (50 mW) to reduce laser hazards during the alignment of the system. The laser power used for alignment should be sufficient enough to be able to visualize the laser beam with an infrared visor and/or with a fluorescent paper plate.
7. Position and align two coated mirrors as in point 3.
8. Position two galvanometric mirrors to control the beam x and y position. Use the previously installed two mirrors and two irises for alignment. The galvo mirrors should be placed at a 45 degree angle.
9. Position two coated lenses to engineer the telescope that will expand the beam. This should be a telescope with convergent lenses (i.e., with positive focal length) to establish complementary planes: the focus of the first lens of the telescope should be complementary to the focus of the objective used to image the sample. Use the magnification formulas to calculate the necessary focal lengths that the lenses needed (refer to **Note 2**). The first lens of the telescope should be placed at a distance equal to its focal length from the galvanometric system. The second lens of the telescope should be positioned after the first lens at a distance equal to the sum of the two focal lengths of the two lenses. To obtain complementary image planes, the second lens should be also positioned at a distance equal to its focal length from the objective. The beam expansion can be performed in two steps with two telescopes to make the system more compact and with a higher degree of safety (refer to **Note 3**).
10. Position the dichroic mirror with 45 degrees close to back aperture of the imaging objective. Make sure the coated region faces the laser beam.

2.2 OPTIC PATH ALIGNMENT ON THE SAMPLE

For a confocal/spinning disc imaging system:

1. Place a slip covered with a layer of fluorescent dye or with a monolayer of fluorescent beads on the stage over the objective.
2. Open the shutter to let the laser beam travel through the optic path that has been set up.
3. The beam enters the objective and is focused on the slip where it will excite the fluorophores or the beads on a spot. Slightly move the first lens of the telescope back and forth on the rails to find the optimal z position: this is reached when the excitation region is reduced to a diffraction limited spot. Displace the dichroic mirror (or the mirror preceding the dichroic, in case the dichroic is mounted on a holder with no degrees of freedom) using the degrees of freedom of the dichroic

mirror holder so that the focused laser beam is centered in the field of view of the imaging CCD camera.

For a light sheet scope with immersion chamber:

1. Fill the scope chamber with PBS (or E3 fish medium) and follow the same procedure as presented for the spinning disc imaging system. The liquid will be excited by the focused infrared laser beam via a two photon process and will emit fluorescent light. The laser power necessary to generate fluorescence in this medium is usually higher than for fluorescent dyes or for fluorescent beads.
2. The alignment is then further optimized by using a *Drosophila* embryo sample with, for instance, membrane tagged by GFP and by focusing on the apical surface of the cells. Reduce the laser power before focusing the laser on the *Drosophila* embryo to avoid major embryo damage during alignment. Gradually increase the laser power until the two-photon excitation spot is visible on the cell apical membrane.

2.3 CONTROL OF THE ELECTROMECHANICAL COMPONENTS

Use LabView software to control the shutter opening and closing, the $\lambda/2$ rotating holder angular position to modulate the laser power, and the galvanometric unit to displace the beam over the sample along the x- and the y-axis. It is to be noticed that displacing the beam along x and y far away from the central focus of the objective (i.e., entering in the back aperture of the objective with great angles) results in a strong degradation of the beam shape and a reduction of the beam intensity. Thus it is recommended to perform laser manipulation experiments in regions not too far away from the central focus of the objective. Instead of displacing the beam via a galvanometric system, it is possible to keep the beam fixed and move the stage on which the sample is positioned. In this case the stage should be controlled via the LabView program to control the position of interaction between the laser beam and the sample. This last procedure does not allow displacing the beam rapidly but allows conserving the beam shape and intensity at the position of interaction with the specimen.

2.4 SAMPLE PREPARATION

For a confocal/spinning disc imaging system:

1. Collect *Drosophila* embryos. Use bleach to degrade the chorion of the embryo and water to wash out the embryo.
2. By using a soft brush, transfer the embryos on a solid substrate of agar gel to keep the dechorionated embryos humid.
3. Select the embryos of interest, align them, and orient them: the region of interest of the embryo should be facing upward if using an inverted scope.

4. Spread a thin layer of heptane glue over a slip.
5. Glue the embryos to the slip by gently pressing the slip on the agar substrate.
6. Cover the embryos with few drops of halocarbon oil.

For a light sheet scope with immersion chamber:

1. Collect *Drosophila* embryos. Use bleach to degrade the chorion and water to wash out the embryo.
2. By using a soft brush, transfer the embryos on a solid substrate of agar gel to keep the dechorionated embryos humid.
3. Choose the embryo of interest and insert it in a capillary filled with agar or other low diffraction gel (Krzic, Gunther, Saunders, Streichan, & Hufnagel, 2012).

2.5 TISSUE CAUTERIZATION PROCEDURE

To perform tissue cauterization, it is necessary to tightly focus the infrared pulsed laser onto the sample. To produce fix boundaries the laser should be focused on a region at the interface between the cell tissue and the vitelline membrane. The laser power necessary to cauterize is lower than to perform plasma-induced ablations; the exposure time necessary to cauterize the cells is longer compared to perform plasma-induced ablations. The procedure to perform tissue cauterization is slightly different when performing this in tandem with a spinning disc or a light sheet (Figs. 1 and 2, respectively) (Refer for this to **Note 4** and to **Note 5**). Perform the experiments at constant temperature (preferably room temperature: e.g., 22°C).

1. Use a *Drosophila* line that allows visualizing the apical region of cells (e.g., membrane marker or junctional marker).
2. Define the region of cauterization by using an ROI tool that will provide the input to the scanning galvanometric mirrors.
3. Tune the laser power by using the attenuator: use 150 mW average power (if using an objective with 1.2 NA) or 200 mW average power (if using an objective of 1.1 NA) measured at the back aperture of the objective. Expose the sample to the infrared femtosecond laser for 40 ms.
4. When performing a cauterization over a line make sure the laser beam is always focused at the interface between the vitelline membrane and the blastoderm tissue to generate a fix boundary (refer to **Note 4** and **Note 6**).

2.6 ELECTRON MICROSCOPY PROCEDURE

Electron microscopy tomograms are taken to inspect with high magnification the cauterized region. This cannot be performed on an embryo that has been glued to a slip since the embryo cannot be easily detached without breaking the vitelline membrane. This protocol is used instead when performing tissue cauterization on an embryo imbedded in gel within a capillary and after, for instance, being imaged with a light sheet microscope. The embryo can be recovered from the gel by using a

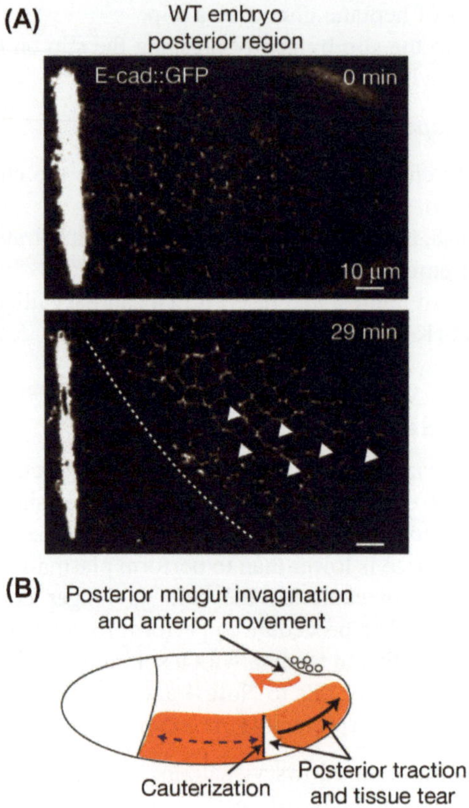

FIGURE 1

Laser cauterization reveals the interaction between the ectoderm and the posterior midgut tissues during *Drosophila* embryo gastrulation. (A) Top panel: laser cauterization is performed before the ectoderm starts to extend. Bottom panel: the tissue tears off from the cauterized region revealing an external pulling force generated by the posterior region of the embryo. Images are taken with a spinning disc imaging unit to which the cauterization system is coupled to. (B) Schematic representing the experiment performed.

Adapted from Collinet, C., Rauzi, M., Lenne, P. F., & Lecuit, T. (2015). Local and tissue-scale forces drive oriented junction growth during tissue extension. Nature Cell Biology, 17, *1247–1258.*

metal needle. Such procedure needs an extreme attention and gentleness since the vitelline membrane of the embryo has been sensitized by the infrared laser. After cauterizing, imaging, and recovering the embryo from the gel cylinder, high-pressure freeze the embryo as soon as possible to be able to relate the last 3-D image of the embryo taken with the light sheet system to the tomograms that will be then taken with the electron microscope (Fig. 3).

1. High-pressure freeze the cauterized embryos (HPM010 AbraFluid), using 20% dextran as cryoprotectant. The embryos should be pierced with a needle in a

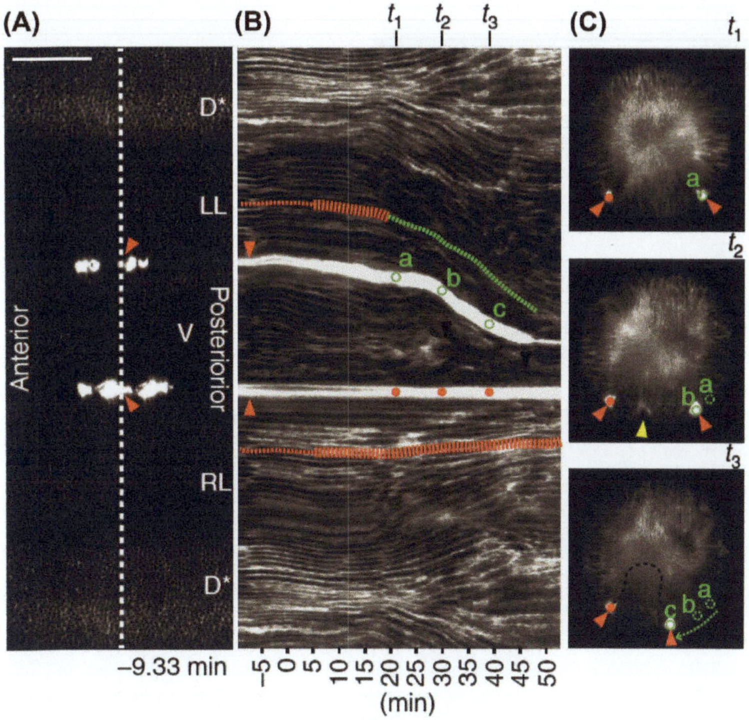

FIGURE 2

Laser cauterization reveals the coordination between the ectoderm and the mesoderm tissues during *Drosophila* embryo gastrulation. (A) Cylindrical projection of the embryo before gastrulation: the dorsal region is shown twice (on the top and on the bottom) to emphasize the symmetry of the system. Two cauterizations (*red arrowheads*) are performed on a line parallel to the anterior–posterior axis of the embryo at the ventral–lateral regions. (B) Kymograph taken in the central region of the embryo along the *dashed line* shown in A. (C) Cross-section views at different time points in the central region of the embryo. Mesoderm internalization is abolished when ectoderm displacement is impaired by laser cauterizations: this shows coordination and interdependency between ectoderm displacement and mesoderm internalization. Images are taken with a multiview light sheet microscope to which the cauterization system is coupled to. Scale bar 100 μm.

Adapted from Rauzi, M., Krzic, U., Saunders, T. E., Krajnc, M., Ziherl, P., Hufnagel, L., & Leptin, M. (2015). Embryo-scale tissue mechanics during Drosophila *gastrulation movements.* Nature Communications, 6.

cryo-microtome chamber (Leica EM FC6) at −160°C to facilitate freeze substitution (Eltsov et al., 2015).

2. Freeze-substituted the embryos (EM-AFS2—Leica Microsystems) with 0.3% uranyl acetate (UA), 0.3% glutaraldehyde and 3% water in acetone at −90°C for 48 h.
3. Raise the temperature to −45°C at 2°C/h and further incubate the samples for 16 h.

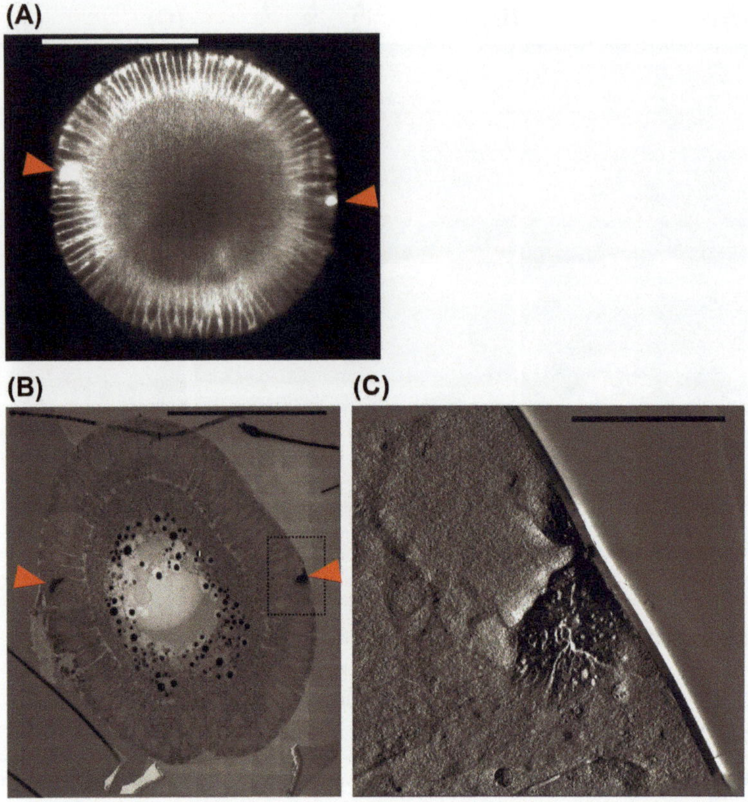

FIGURE 3

Electron microscope tomograms of the cauterized regions. (A) Cross-section view of a living membrane-tagged embryo on which two cauterizations are performed: one at the surface (*left arrowhead*) and one 5 μm inside the epithelium (*right arrowhead*) before embryo gastrulation. Scale bar 100 μm. (B) Electron microscope tomogram of a cross section of the embryo shown in (A) few minutes afterward. Laser cauterization locally increases the electron density of the tissue (see *arrowheads*). Scale bar 100 μm. (C) Electron microscope tomogram showing a close-up view of the region delimited by a *dashed line* in B. Scale bar 1 μm.

4. Rinse the samples in acetone.
5. Infiltrate the samples in Lowicryl HM20 resin, while raising the temperature to −25°C and leave the samples polymerize under UV light for 48 h at −25°C and for further 9 h while the temperature is gradually raised to 20°C (5°C/h).
6. Cut thin (70 nm) and/or thick sections (300 nm) from the polymerized resin block in which the embryo is embedded.
7. Collect the sections on Formvar-coated slot grids.
8. Acquire 2-D TEM images with a FEI CM120 electron microscope.

9. For tomography, acquire tilt series with a FEI Tecnai F30 electron microscope and reconstruct the tomograms using the IMOD software package (Kremer, Mastronarde, & McIntosh, 1996).

2.7 SAFETY PROCEDURE

Since the laser used for cauterizations is a class 4 laser, (thus with the highest degree of danger) make sure to follow the safety procedure to avoid eventual hazards:

1. Set up the optical path as close as possible to the optic table.
2. The room where experiments are performed should have controlled access.
3. Use adapted eye protecting goggles while setting up the system and while using them for experiments if the system is not intubated.
4. Work in bright room conditions as much as possible to reduce the NA of the eye that reduces the risk of eye hazard.
5. Avoid wearing bracelets or watches while setting up the system to avoid any possible reflections.
6. Intubate the system once it is built to reduce the risk of eventual skin or eye hazards.

Refer to the European (EN 207/208/60825) and American (ANSI Z136) safety norms.

2.8 NOTES

1. Position, as far as possible, the first iris after the first and before the second lens and the second iris after the second lens. Move the beam into the first iris using the degrees of freedom of the first mirror holder and into the second iris using the degrees of freedom of the second mirror holder. Repeat this procedure in cycles until reaching optimal alignment.
2. If we consider "d" being the laser beam width ($1/e^2$) and "D" the diameter of the back aperture of the objective, an expansion factor of $D/d = M$ is needed. Select two plano-convex lenses with focal lengths f1 (for the first lens) and f2 (for the second lens positioned closer to the imaging objective) so that $f2/f1 = M$. Remember to place the convex side of the lenses towards the outside of the telescope where the beam is supposed to be parallel. Avoid using lenses with focal length inferior to 30 mm to avoid degrading the beam quality.
3. If a preexpansion of the beam is envisioned to reduce the size of the overall system this can be performed just before the galvanometric mirror apparatus. Since for this preexpansion there is no conjugated plane condition to satisfy, it is preferable to engineer a telescope with one divergent lens and one convergent lens. In this way the beam is not focused allowing to maintain a higher degree of safety (since the intensity of the laser beam is not increased) and to avoid generating eventual plasmas in the air. In this case, lenses should be placed at a distance equal to $|f2|-|f1|$, one from the other.

4. Tissue cauterization with boundary fixation is more easily achieved on a sample mounted for spinning disc imaging (Collinet et al., 2015) than on a specimen imbedded in a gel within a capillary for light sheet imaging (Rauzi et al., 2015). The embryo in fact presents a flatter surface in the first case given the slight pressure used to glue it to the slip (refer to the Section 2.4). When the sample is mounted in a gel cylinder, it maintains its curvature. When performing a cauterization line is then necessary to correct the z position to maintain the laser beam focused on the interface between the blastoderm cells and the vitelline membrane.
5. When mounting the sample on a capillary this can usually be rotated. On a light sheet system with immersion chamber is thus possible to perform several cauterizations on different regions of the embryo (Rauzi et al., 2015).
6. Tissue cauterization can be performed at different tissue depths by exploiting the infrared multiphoton feature of the process. If the cauterization is performed at the interface between the tissue and the vitelline membrane, this causes the tissue to anchor and fuse to the vitelline membrane. Given the fact that the vitelline membrane is not mobile, this allows to generate fix boundaries. If the cauterization is performed at the very surface of the vitelline membrane, this can cause membrane piercing and eventually cell and yolk mass leakages. If the cauterization is performed in the bulk (e.g., 10 μm inside the embryo) this will not generate fix boundaries. Finally, if the cauterization is performed 2–5 μm below the interface between the vitelline membrane and the cells, this can generate transient fix boundaries that will impair tissue displacement but that can eventually detach restoring tissue dynamics.

CONCLUSION

Laser-induced tissue cauterization can generate fix boundaries in the developing *Drosophila* embryo. Thus, such technique allows performing unprecedented experiments in which, for instance, global cellular flows can be impaired during embryo development with very high spatial and temporal specificity. With this technique the apical layer of the blastoderm epithelium stitches to the vitelline membrane that acts as immobile scaffold containing the embryo. This technique is now part of a new toolkit that scientists can use to probe tissue interaction, tissue coordination, and embryo scale mechanics.

ACKNOWLEDGMENTS

I thank Paolo Ronchi from the EMBL Electron Microscopy Core Facility for providing the electron tomograms. I thank the laser company Amplitude Systèmes for continuous support. This work would have not been possible without the contribution of Lars Hufnagel, Maria Leptin, and Uros Krzic. I also thank Gustavo Quintas Glasner de Medeiros, Dimitri Kromm, and Balint Balazs for their work to further develop the technique. Matteo Rauzi is supported by the Université Nice Sophia Antipolis, by the Institut de Biologie Valrose, by the Mission interdisciplinarité PEPS CNRS and by the Actions Recherche UCA Jedi.

REFERENCES

Cavey, M., & Lecuit, T. (2008). Imaging cellular and molecular dynamics in live embryos using fluorescent proteins. *Methods in Molecular Biology, 420*, 219–238.

Collinet, C., Rauzi, M., Lenne, P. F., & Lecuit, T. (2015). Local and tissue-scale forces drive oriented junction growth during tissue extension. *Nature Cell Biology, 17*, 1247–1258.

Eltsov, M., Dube, N., Yu, Z., Pasakarnis, L., Haselmann-Weiss, U., Brunner, D., & Frangakis, A. S. (2015). Quantitative analysis of cytoskeletal reorganization during epithelial tissue sealing by large-volume electron tomography. *Nature Cell Biology, 17*, 605–614.

Kremer, J. R., Mastronarde, D. N., & McIntosh, J. R. (1996). Computer visualization of three-dimensional image data using IMOD. *Journal of Structural Biology, 116*, 71–76.

Krzic, U., Gunther, S., Saunders, T. E., Streichan, S. J., & Hufnagel, L. (2012). Multiview light-sheet microscope for rapid in toto imaging. *Nature Methods, 9*, 730–733.

Rauzi, M., Krzic, U., Saunders, T. E., Krajnc, M., Ziherl, P., Hufnagel, L., & Leptin, M. (2015). Embryo-scale tissue mechanics during *Drosophila* gastrulation movements. *Nature Communications, 6*.

Waring, G. L. (2000). Morphogenesis of the eggshell in *Drosophila*. *International Review of Cytology, 198*, 67–108.

CHAPTER

9

Optogenetic inhibition of apical constriction during *Drosophila* embryonic development

G. Guglielmi, S. De Renzis[1]
European Molecular Biology Laboratory (EMBL), Heidelberg, Germany
[1]*Corresponding author: E-mail: stefano.derenzis@embl.de*

CHAPTER OUTLINE

Abstract

Morphogenesis of multicellular organisms is driven by changes in cell behavior, which happen at precise locations and defined developmental stages. Therefore, the studying of morphogenetic events would greatly benefit from tools that allow the perturbation of cell activity with spatial and temporal precision. We recently developed an optogenetic approach to modulate cell contractility with cellular precision and on fast (seconds) timescales during *Drosophila* embryogenesis. We present here a protocol to handle genetically engineered photosensitive *Drosophila* embryos and achieve light-mediated

Methods in Cell Biology, Volume 139, ISSN 0091-679X, http://dx.doi.org/10.1016/bs.mcb.2016.10.007

inhibition of apical constriction during tissue invagination. The possibility to modulate the levels of optogenetic activation at different laser powers makes this method suited also for studying how mechanical stresses are sensed and interpreted in vivo. Given the conserved function of cell contractility during animal development, the application of this method to other morphogenetic processes will facilitate our understanding of tissue mechanics and cell—cell interaction during morphogenesis.

INTRODUCTION

Animal morphogenesis is characterized by a set of conserved molecular and cellular behaviors, including contractility, adhesion, signaling, and proliferation, which change as a function of time and space. Traditionally, morphogenesis has been studied by broadly interfering with protein activity, for example, through standard genetic tools or chemical perturbations, and then examining the effects of such perturbations on embryonic development. However, chemical and genetic perturbations do not allow the perturbation of biological systems with the desirable spatial and temporal precision, which ideally should be on the second timescale and at the cellular/subcellular level.

Let us consider cell contractility, an actomyosin-driven process that is key for many morphogenetic events, including tissue invagination, cell migration, and convergent extension (reviewed in Munjal & Lecuit, 2014). At the individual cell level, spatially localized actomyosin-driven cortical contraction results in dramatic changes in cell shape, motility, or junction stability. At the tissue level, the forces generated by the actomyosin cytoskeleton are transmitted between neighboring cells and between cells and the extracellular matrix via cadherin- or integrin-based adhesive structures (reviewed in Barone & Heisenberg, 2012; Lecuit, Lenne, & Munro, 2011), thus causing changes in morphology on a larger scale. Some of the questions that remain difficult to address with genetic or chemical tools relate to (1) when and where, within a cell or a tissue, actomyosin-driven contractility is required; (2) how cell autonomous changes in cell contractility influence adjacent cells and global physical restructuring; and (3) how tissue features, such as tension or geometry, influence individual cell behavior. In the last decade, laser ablation—by enabling the severing of actomyosin-containing structures at specific junctional interfaces—has emerged as a useful tool to start elucidating some of these aspects (Hutson et al., 2003; Martin, Gelbart, Fernandez-Gonzalez, Kaschube, & Wieschaus, 2010; Rauzi, Verant, Lecuit, & Lenne, 2008; Solon, Kaya-Copur, Colombelli, & Brunner, 2009). However, the potential phototoxicity induced by high-intensity ultraviolet or near-infrared lasers has limited the investigation of long timescale responses (Colombelli & Solon, 2013). More recently, optogenetics—by offering the possibility to control protein activity or location with light—has proved to be a powerful technique to modulate cell behavior with both spatial and temporal precision (reviewed in Toettcher, Voigt, Weiner, & Lim, 2011). Optogenetic modulation of small GTPases has enabled researchers to define the temporal dynamics of Ras/ERK signaling in individual cultured cells (Toettcher, Weiner, & Lim, 2013), and the spatial requirements of Rac activity during the migration of cell collectives (Cai et al., 2014; Wang, He, Wu, Hahn, & Montell, 2010).

We took advantage of the optogenetic toolkit to develop an approach to control cell contractility in intact *Drosophila* embryos (Guglielmi, Barry, Huber, & De Renzis, 2015). To do so, we decided to target $PI(4,5)P_2$, a plasma membrane phosphoinositide that acts as a scaffold for several actin regulators (reviewed in Saarikangas, Zhao, & Lappalainen, 2010) and that has been shown to be essential for cell morphogenesis during early *Drosophila* development (Reversi, Loeser, Subramanian, Schultz, & De Renzis, 2014). We employed the Cry2-CIB1 light-sensitive modules to target the catalytic domain of the inositol-polyphosphate-5-phosphatase OCRL (5-ptase_{OCRL}) (Idevall-Hagren, Dickson, Hille, Toomre, & De Camilli, 2012; Kennedy et al., 2010) to the plasma membrane of *Drosophila* embryos. We generated transgenic embryos where the GFP-tagged N-terminal domain of CIB1 (CIBN::pmGFP) localizes to the plasma membrane, and the mCherry-tagged PHR domain of Cry2 is fused to the 5-ptase ($\text{mCh::Cry2::5-ptase}_{OCRL}$). In the absence of blue light (488 nm), Cry2 and CIB1 do not interact, therefore the 5-ptase remains localized in the cytosol. On irradiation with a 488 nm laser and subsequent interaction between Cry2 and CIB1, the 5-ptase relocalizes to the plasma membrane and depletes $PI(4,5)P_2$ by dephosphorylating it to PI(4)P. This in turn results in actin depletion, blockade of cell contractility, and inhibition of morphogenetic movements (Fig. 1).

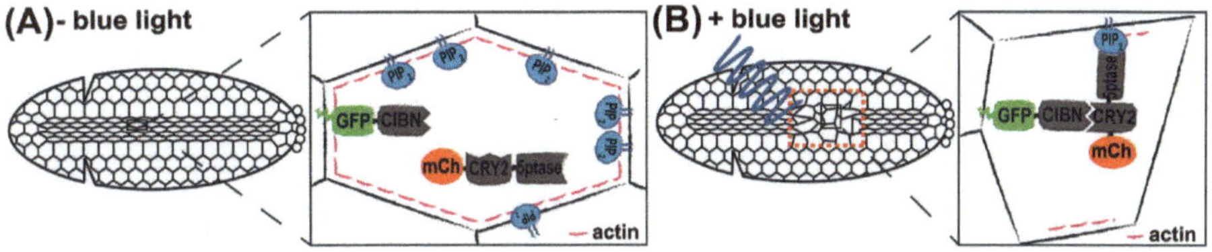

FIGURE 1 An optogenetic approach to achieve local regulation of apical constriction during tissue morphogenesis.

(A and B) Schematic drawing of a surface view of the *Drosophila* mesoderm during ventral furrow formation. GFP-tagged CIBN localizes to the plasma membrane. Cry2 is tagged with mCherry and fused to the catalytic domain of the 5-ptase. (A) In the absence of blue light, the 5-ptase is cytosolic and morphogenesis progresses normally. (B) On blue light illumination, CIBN and Cry2 interact, causing the recruitment of the 5-ptase to the plasma membrane. This results in $PI(4,5)P_2$ and, in turn, cortical actin depletion, which leads to the inhibition of apical constriction.

Panels adapted from Guglielmi, G., Barry, J.D., Huber, W., & De Renzis, S. (2015). An optogenetic method to modulate cell contractility during tissue morphogenesis. Developmental Cell. *http://dx.doi.org/10.1016/j.devcel.2015.10.020, http://www.sciencedirect.com/science/article/pii/S1534580715006851 (http://dx.doi.org/10.1016/j.devcel.2015.10.020), published under a Creative Commons Attribution License (CC BY): http://creativecommons.org/licenses/by/4.0/.*

We could show that a light input at $\lambda = 488$ or 950 nm (two-photon mode) resulted in the translocation of mCh::Cry2::5-ptase$_{OCRL}$ to the plasma membrane on short (seconds) timescales, and that blue light–mediated relocalization of mCh::Cry2::5-ptase$_{OCRL}$ caused PI(4,5)P_2 and actin depletion from the plasma membrane. Consistently, we could further show that optogenetic activation results in the arrest of two well-established actin-dependent morphogenetic processes, the cellularization of the syncytial blastoderm and the internalization of the presumptive mesoderm through ventral furrow formation.

In particular, we focused on ventral furrow formation, the first large-scale tissue movement during *Drosophila* embryogenesis that has served for many years as a model for studying tissue invagination. At around 4 h postfertilization, a group of ~1000 epithelial cells aligned along the embryonic anterior–posterior (a-p) axis on the ventral surface of the embryo contract at their apices and invaginate (Leptin & Grunewald, 1990; Sweeton, Parks, Costa, & Wieschaus, 1991). Contractility in ventral cells is driven by the recruitment and activation of Myosin II and reorganization of actin filaments at the apical cortex (apical constriction). The contractile activity of the apical actomyosin network acts as a force generator causing shrinkage of the apical surface and consequent cell shape changes (Martin, Kaschube, & Wieschaus, 2009; Mason, Tworoger, & Martin, 2013). The propagation of forces across the tissue through cadherin-based adherens junctions results in the generation of a tissue-level tension that causes the bending of the entire epithelial sheet (Martin et al., 2010). Experimental evidences showing that forces generated in ventral cells drive tissue invagination are compelling although indirect. Furthermore, whether apical constriction is required only during the initial stages of ventral furrow formation, as suggested by computational models (Spahn & Reuter, 2013), or throughout the invagination process was unknown. By locally inhibiting apical constriction in ventral cells, we could show that forces generated in ventral cells are required for invagination and are necessary throughout the process of ventral furrow formation.

Another open question is related to the impact of tissue geometry on the contractile behavior of individual cells within the ventral furrow primordium. Ventral cells constrict preferentially along the embryonic dorsal–ventral (d-v) axis, and elongate along its anterior–posterior (a-p) axis. This asymmetric constriction, known as a-p anisotropic constriction, is thought to be caused by tissue forces that are higher along the a-p axis than along the d-v axis. Local modulation of apical constriction enabled us to show that local geometrical constraints determine the way individual cells constrict. Individual cells constrict in a more anisotropic way if they are organized within a rectangular-shaped tissue than if they are contained in a square-shaped tissue.

Here, we describe a detailed protocol to generate and handle photosensitive *Drosophila* embryos and achieve optogenetic inhibition of apical constriction during gastrulation.

1. METHODS

1.1 MOLECULAR BIOLOGY

All constructs were cloned according to the manufacturer's guidelines using the Gateway cloning system (Invitrogen by Life Technologies) into pPW, a Gateway vector containing a UASp cassette optimized for expression during oogenesis (Rorth, 1998).

The plasma membrane anchor CIBN::pmGFP was amplified from pCIBN(deltaNLS)-pmGFP. This plasmid was generated by Kennedy et al. (2010) by cloning the N-terminal end of CIB1 (amino acids 1-170) into a version of pEGFP-C1 (Clontech) containing the CaaX sequence from Kras4B at the C-terminal end. We chose to use the truncated version of CIB1 as it is showed lower dark−state activity (i.e., interaction with Cry2 in the absence of blue light) compared to the full-length CIB1. Furthermore, CIBN protein lacks its nuclear localization sequence, thus ensuring specific plasma membrane localization (Kennedy et al., 2010).

We chose to use only the photolyase homology region (PHR) of Cry2 (amino acids 1−498), since this is the domain responsible for light sensitivity, and it has been previously shown to result in higher expression and reporter activation in cell culture compared to the full-length Cry2 (Kennedy et al., 2010). We amplified the PHR domain of Cry2 from pCRY2PHR-mCherryN1 generated by Kennedy et al. (2010), and fused it downstream of the sequence encoding monomeric (m)Cherry. The sequences encoding the PHR domain and mCherry were spaced by a GAGA linker to allow flexibility between the fluorescent protein and the photosensitive domain. The catalytic domain (amino acids 99−519) of *Drosophila melanogaster* inositol-polyphosphate 5-phosphatase OCRL (dOCRL) was amplified from *D. melanogaster* cDNA using gene-specific primers and fused downstream of the mCh::Cry2 construct through overlap PCR.

Table 1 shows the primers used to generate all constructs. Primers were purchased from Sigma−Aldrich. On arrival, lyophilized primers were resuspended in sterile ddH_2O.

1.1.1 Notes

Instead of CIBN, one may opt to tether the photosensitive module ($Cry2_{PHR}$) to the plasma membrane by fusing it to a CaaX sequence. However, this strategy has proved ineffective, as the plasma membrane−anchored Cry2 and the cytosolic CIBN interacted at much lower levels when compared to the membrane-anchored CIBN and the cytosolic Cry2 (Idevall-Hagren et al., 2012).

1.1.2 Materials

- pPW vector (Drosophila Genomics Resource Center, Bloomington, IN—stock no. 1130)

Table 1 List of Primers Used to Generate the Plasma Membrane Anchor CIBN::pmGFP and mCh::Cry2::5-ptase$_{OCRL}$

Name	Sequence (5′→3′)	Comments
CIBN F	*CACCATGAATGGAGCTATAGGAGG*	CIBN forward + 5′ CACC sequence for cloning into pENTR
CIBN R CAAX	*CTTCTTTTTCTTCTTTTTACCCTTATACAGATGAATATAATCCGTTTTCTC*	EGFP-CaaX reverse
mCherry F	*CACCATGGTGACGAAGGGCGAGGAGGATAA*	mCherry forward + 5′ CACC sequence for cloning into pENTR
mChGAGA R (Cry2)	*ATCTTCATGGCGCCGGCGCCCTTGTACAG*	mCherry-GAGA linker reverse + overhang with PHR domain of Cry2
Cry2PHR F (mChGAGA)	*CGCGCGGCGCCATGAAGATGGACAAAAAG*	PHR domain forward + overhang with mCherry-GAGA linker
Cry2PHR R (dOCRL)	*AGCTCTTGCTGTCTGCTGCTCCGATCATGATC*	PHR domain reverse + overhang with dOCRL
dOCRL F (Cry2)	*AGCAGCAGACAGCAAGAGCTCGATCACCAC*	dOCRL forward + overhang with PHR domain of Cry2
dOCRL R	*TCAGAAATCGATGACCGTCTTCTCG*	dOCRL reverse + STOP codon

- pENTR/D-TOPO Cloning Kit (Invitrogen by Life Technologies—catalog no. K240020)
- Gateway LR Clonase II Plus Enzyme Mix (Invitrogen by Life Technologies—catalog no. 12538120)
- pCIBN(deltaNLS)-pmGFP (Addgene—plasmid no. 26867)
- pCRY2PHR-mCherryN1 (Addgene—plasmid no. 26866)
- Forward and reverse primers (Sigma–Aldrich), see Table 1

1.2 DROSOPHILA GENETICS

Optogenetic inhibition of apical constriction can be achieved by illuminating with blue light embryos coexpressing the plasma membrane anchor CIBN::pmGFP and the catalytic domain of the 5-ptase OCRL fused to the PHR domain of Cry2 and tagged with mCherry (mCh::Cry2::5-ptase$_{OCRL}$). The GFP and mCherry tags are dispensable and are used only to visualize the plasma membrane and verify the efficiency of recruitment, respectively. Using standard P element-mediated transformation, we generated two stable transgenic stocks, available on request that carry either CIBN::pmGFP or mCh::Cry2::5-ptase$_{OCRL}$ under the UASp promoter. The UASp sequence was chosen to allow for efficient expression of the transgenes during oogenesis, so that embryos would be loaded with the desired proteins during early developmental stages.

Embryos coexpressing CIBN::pmGFP or mCh::Cry2::5-ptase$_{OCRL}$ are generated as follows:

1. Virgin females balanced on the third chromosome and carrying the CIBN::pmGFP transgene on the second chromosome are crossed to males balanced on the second chromosome and carrying the mCh::Cry2::5-ptase$_{OCRL}$ transgene on the third chromosome:

$$\text{♀ w/w; CIBN::pmGFP/CyO; Sb/TM3, Ser}$$
$$\text{(x)}$$
$$\text{♂ w/<; If/CyO; mCh::Cry2::5-ptase}_{\text{OCRL}}\text{/TM3, Ser}$$

2. Immediately after setting up the cross, the vial is transferred to a fiberboard box (Fig. 2A) and kept at 22°C to avoid any unwanted activation in case of leaky expression of the transgenes. Once the progeny hatches, males of the right genotype are selected and crossed to virgin females expressing Gal4 under the maternal tubulin promoter:

$$\text{♂ w/<; CIBN::pmGFP/CyO; mCh::Cry2::5-ptase}_{\text{OCRL}}\text{/TM3, Ser}$$
$$\text{(x)}$$
$$\text{♀ w/w; Gal4/Gal4; Gal4/Gal4}$$

3. The vial containing the new cross is transferred to a fiberboard box and kept at 22°C until the progeny hatches. At this point, male and female flies of the right

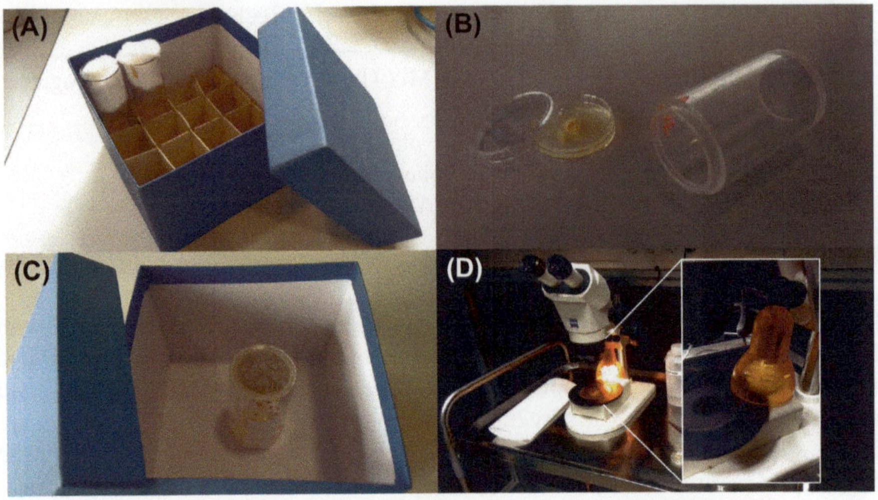

FIGURE 2 A step-by-step guide to handle photosensitive *Drosophila* embryos.

(A) Vials containing photosensitive flies are kept in fiberboard boxes. (B) Flies of genotype (w; CIBN::pmGFP/Gal4; mCh::Cry2::5-ptase$_{OCRL}$ /Gal4) are selected and transferred to an embryo collection cage (right) which is covered with a fresh apple juice plate smeared with yeast paste (left). (C) The embryo collection cage is kept in a fiberboard box at 18°C. (D) In a dark room, embryos are visualized under a dissection microscope in which the source of light has been replaced with a lamp equipped with a yellow bulb (inset).

genotype (w; CIBN::pmGFP/Gal4; mCh::Cry2::5-ptase$_{OCRL}$/Gal4) are selected and transferred to an embryo collection cage (Fig. 2B, right) which is then covered with a fresh apple juice plate smeared with half teaspoon of yeast paste (Fig. 2B, left) and kept in a fiberboard box at 18°C (Fig. 2C).

1.2.1 Notes

Although a stable stock of genotype (w; CIBN::pmGFP/CyO; mCh::Cry2::5-ptase$_{OCRL}$/TM3, Ser) is viable and fertile, we observed a decrease in expression of the two transgenes over time. Therefore, we recommend starting from the individual stocks to ensure optimal expression of both CIBN::pmGFP and mCh::Cry2::5-ptase$_{OCRL}$.

1.2.2 Materials

- *Drosophila melanogaster* stocks:
  - w[*]; P[w+, matαTub>Gal4::VP16]; [w+, mat αTub>Gal4::VP16] (E. Wieschaus, Princeton University, Princeton, NJ). The sequence encoding Gal4::VP16 was cloned downstream of the maternal tubulin promoter into a CaSpeR-based vector.
  - w[*]; P[w+, UASp>CIBN::pmGFP]/CyO; Sb/TM3, Ser (S. De Renzis, European Molecular Biology Laboratory, Heidelberg, Germany). The

  GFP-tagged N-terminal domain of CIB1 was fused to a CaaX anchor and cloned into pPW vector (Drosophila Genomics Resource Center, Bloomington, IN).
  - w[*]; If/CyO; P[w+, UASp>mCherry::Cry2$_{PHR}$::5-ptase$_{OCRL}$]/TM3, Ser (S. De Renzis, European Molecular Biology Laboratory, Heidelberg, Germany). The mCherry-tagged PHR domain of Cry2 was cloned upstream of the *Drosophila* inositol-polyphosphate 5-phosphatase OCRL into pPW vector (Drosophila Genomics Resource Center, Bloomington, IN).
- Fiberboard box 136 × 136 × 100 mm (Alpha Box 100, National Lab—catalog number KA1100BU1)
- Embryo collection cages (EMBL mechanical workshop, see Fig. 2B, right)
- Apple juice plates (recipe for 1L):
  - 22.5 g agar
  - 25 g sucrose
  - 1.5 g methyl-4-hydroxybenzoate
  - 250 mL apple juice
  - 750 mL sterile H_2O
  - pour into 35 × 10 mm dishes (Falcon), let it cool down and store at 4°C

1.3 EMBRYO COLLECTION AND MOUNTING

Ventral furrow forms at around 4 h after fertilization (at 22°C), or stage 6 according to the classification of (Campos-Ortega & Hartenstein, 1997). At 3 h postfertilization, or stage 5, the *Drosophila* embryo is a syncytium made of ~6000 cortically anchored nuclei. At this stage, the apical plasma membrane starts to invaginate between the nuclei to give rise to a single-layered epithelium of approximately 6000 cells. On completion of this process (~50 min at 22°C), known as cellularization, ventral furrow formation begins as a rectangular stripe of ~1000 cells aligned along the a-p axis on the ventral side of the embryo start to constrict their apices (apical constriction) and invaginate. The entire process, from the onset of apical constriction to the end of internalization, lasts for ~10–15 min at 22°C. Therefore, to be able to perturb apical constriction as it arises, it is important to select cellularizing embryos, which can be mounted in the dark and photoactivated at the onset of apical constrictions in ventral cells.

Light-sensitive embryos are collected and mounted for imaging/photoactivation in the following way:

1. A fresh apple juice plate is spread with yeast paste and, in a dark room, put onto the embryo collection cage, which is then transferred back to the fiberboard box and kept at 18°C. After 5–6 h, embryos at the cellularization stage (stage 5) are ready to be collected.
2. In a dark room, the apple juice plate is collected and the yeast paste is removed with a cotton swab. The plate is covered with a thin layer of halocarbon oil 27 and embryos are visualized under an upright dissection microscope equipped

with a 10× objective. To avoid unwanted photoactivation, the microscope's light source is replaced with a lamp equipped with a yellow bulb (Fig. 2D).

3. Cellularizing embryos are selected and gently collected using tweezers. Under oil, stage 5 embryos display a dark yolky core surrounded by a cortical layer of clear cytoplasm, a feature that makes them easily identifiable.
4. To remove the chorion, staged embryos are placed on a small (~2 cm^2) piece of filter paper or kitchen roll soaked in 100% sodium hypochlorite (Fig. 3A).
5. After 1 min, the filter paper is washed three times with dH_2O (Fig. 3B) and then dabbed on another piece of filter paper or kitchen roll to dry out (Fig. 3C).
6. With the help of a probe, embryos are carefully transferred to a 35-mm glass bottom microscopy dish and gently turned so that their ventral side is in contact with the glass bottom (Fig. 3D).
7. The microscopy dish is covered with 2 mL of fresh PBS.

1.3.1 Notes

Embryo collection and mounting should be done strictly in the dark. Even dim ambient light can result in unwanted photoactivation. To have a safe source of light, it is possible to replace white bulbs with yellow ones, or cover light sources with amber lighting filters (see Fig. 3E and next section).

Avoid keeping the embryos in bleach more than 1 min or letting them dry out for too long, as this may damage them.

1.3.2 Materials

- Apple juice plates (see Section 2.2)
- Yeast paste: dry yeast (Saf-Instant, Lesaffre—catalog no. 15909) + dH_2O
- Dissection microscope (Stemi, 2000; Carl Zeiss), equipped with a 10× objective (Carl Zeiss)
- Halocarbon oil 27 (Sigma–Aldrich—catalog no. H8773)
- Clip desk lamp, orange (House Addition—catalog no. QGL1332)
- Décor Color Yellow 11W incandescence bulb (Osram—EAN code 4008321480910)
- 35 mm glass bottom dish, 14 mm microwell, coverslip thickness: 0.16–0.19 mm (MatTek Corporation—catalog no. P35G-1.5-14-C)
- 100% sodium hypochlorite solution (Merck Millipore—catalog no. 105614)
- PBS 1×, pH: 7.4 (EMBL media kitchen)
- dH_2O
- Filter paper 90 cm ∅ (Whatman—catalog no. 1450-090) or kitchen roll (Wischfix, Fripa—catalog no. 95,658)
- Inox tweezers (Dumont—catalog no. 0108-5-PO)

1.4 LIVE IMAGING AND PHOTOACTIVATION

One of the biggest advantages of optogenetics is that light can be used to perturb and observe cell behavior at the same time. In the optogenetic approach we developed,

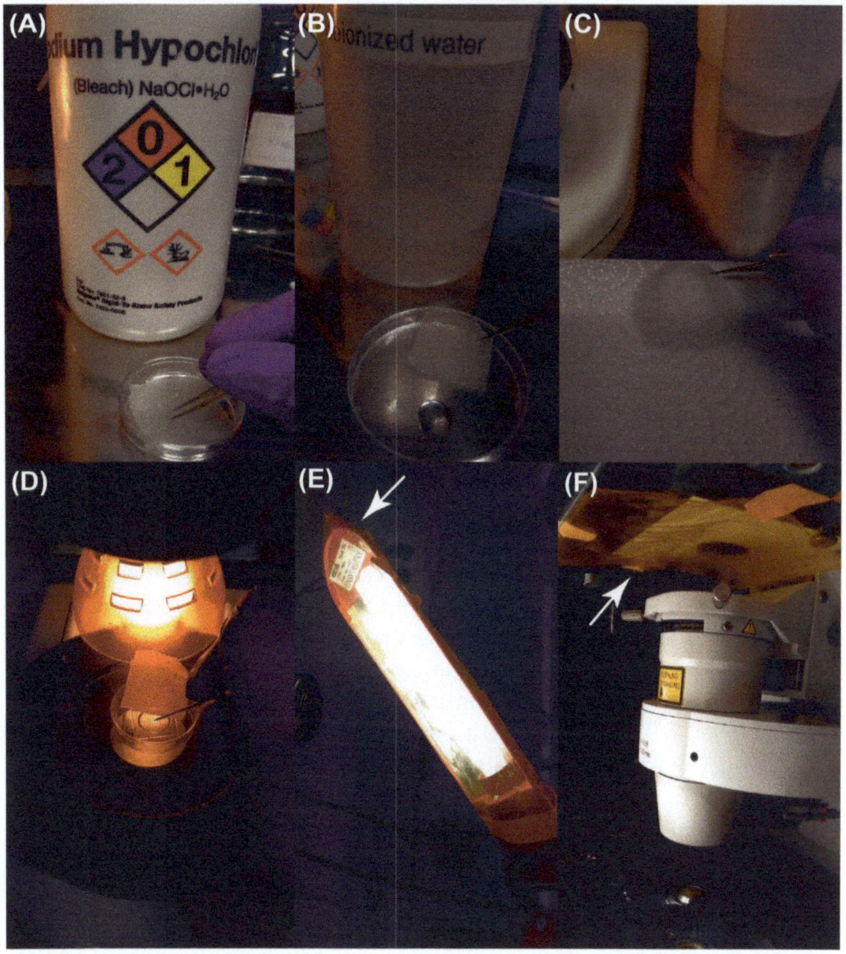

FIGURE 3 An illustrated protocol to mount photosensitive embryos for imaging and photoactivation.

(A) In a dark room, staged embryos are collected with tweezers and dechorionated by placing them on a piece of absorbent paper soaked in 100% sodium hypochlorite. (B) After 1 min, the absorbent paper is washed three times in dH_2O to remove the sodium hypochlorite. (C) The absorbent paper is dabbed on another piece of absorbent paper to dry out. (D) The embryos are carefully mounted on a microscopy dish and turned so that their ventral side is in contact with the glass bottom. (E) Light sources and (F) bright field illumination should be shielded with amber lighting filters (*white arrows*) to avoid unwanted photoactivation.

the plasma membrane anchor is tagged with GFP, and the fusion protein Cry2::5-ptase$_{OCRL}$ is tagged with mCherry (mCh::Cry2::5-ptase$_{OCRL}$). The fluorescent tags are key to observe protein localization in vivo: by monitoring, for instance, the localization of mCherry, we could follow the translocation of the 5-ptase to the plasma membrane (which can also be visualized during photoactivation, if CIBN::pmGFP is used), and determine whether photoactivation has been effective. In addition, quantitative data can be obtained by calculating the average plasma membrane versus cytosol intensity of mCh::Cry2::5-ptase$_{OCRL}$, which is a measure of the photoactivation levels.

Live imaging of embryos is achieved using a 561 nm laser, which excites mCherry and does not cause photoactivation. Actual photoactivation can be obtained with either one-photon (458–488 nm) or two-photon (900–950 nm) illumination, as described in the following sections:

1. In a dark room, the microscopy dish is placed on an inverted confocal laser-scanning microscope equipped with a 561 nm diode laser, an argon laser, and a two-photon laser. To avoid unwanted photoactivation, the transmitted light source is shielded with an amber lighting filter (Fig. 3F).
2. Embryos are visualized with a C-Apochromat 63x/NA 1.2 water immersion objective using a 561 nm laser (laser power: 50.8 μW, measured 1 cm from the objective using a laser power meter).

3a. Photoactivation with one-photon illumination can be achieved using light at $\lambda = 458$, 488, and 514 nm. 488 nm has proved the most effective wavelength for triggering the translocation of mCh::Cry2::5-ptase$_{OCRL}$ to the plasma membrane, where CIBN::pmGFP is anchored (Fig. 4A–D). One-photon activation is achieved by illuminating a single z-stack 5 μm from the apical surface of the embryo using a continuous wave argon laser at $\lambda = 488$ nm (laser power: 6.9 μW, measured 1 cm from the objective) at a scanning speed of 1×10^{-6} s/pixel and averaging of four in bidirectional mode, which corresponds to a scan time of ~2.1 ms per cell and 1 s for the entire embryo. Two consecutive photoactivations are intervalled with a cycle of imaging of mCh::Cry2::5-ptase$_{OCRL}$ at 561 nm, followed by a 30 s dark interval. To inhibit apical constriction, photoactivation is started ~5 min before ventral cells would normally start to constrict. In contrast to embryos expressing only the plasma membrane anchor CIBN::pmGFP, which display normal ventral furrow dynamics (Fig. 4E–G), embryos coexpressing CIBN::pmGFP and mCh::Cry2$_{PHR}$::5-ptase$_{OCRL}$ that are globally photoactivated fail to form a ventral furrow, as showed in Fig. 4H–J.
Although one-photon illumination is very efficient in causing photoactivation and inhibition of apical constriction, it does not allow the achievement of localized patterns of photoactivation due to light scattering. To overcome this limitation we developed a two photon–based photoactivation protocol, which allowed us to obtain efficient and precise photoactivation with cellular precision. Indeed, since this process requires the concomitant absorption of two photons, the excitation of light-sensitive molecules outside of the focal area is extremely unlikely. However, the lower probability of simultaneous

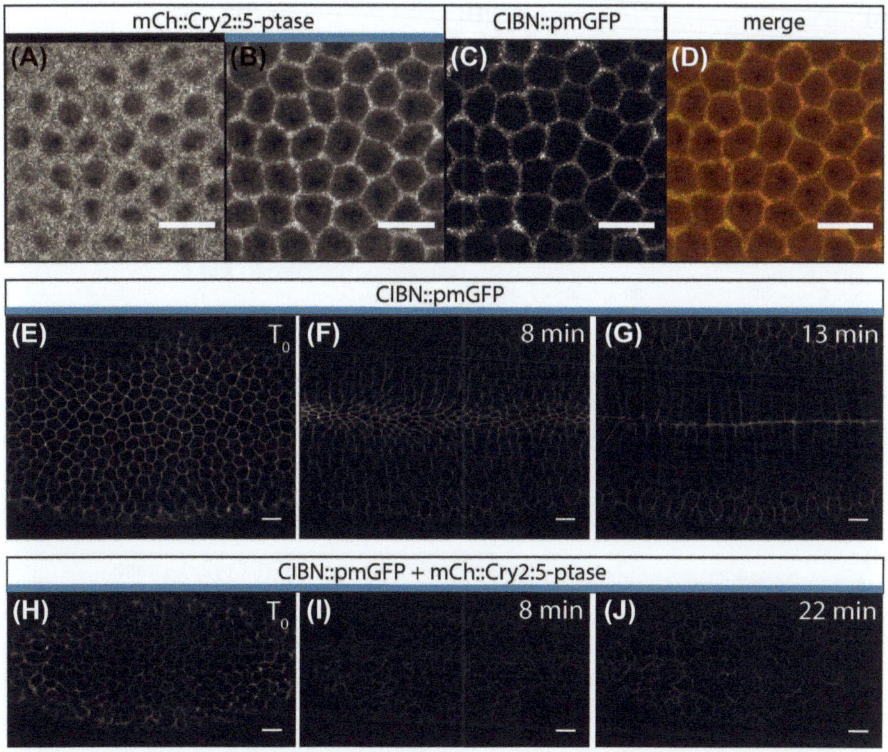

FIGURE 4 Optogenetic activation using one-photon illumination ensures rapid plasma membrane recruitment of mCh::Cry2::5-ptase$_{OCRL}$ and blockade of ventral furrow invagination.

(A–D) Top view of an embryo at the blastoderm stage coexpressing mCh::Cry2::5-ptase$_{OCRL}$ and CIBN::pmGFP. (A) In the absence of blue light, mCh::Cry2::5-ptase$_{OCRL}$ is cytosolic. (B) On a 1 s pulse of 488 nm light, mCh::Cry2::5-ptase$_{OCRL}$ relocalizes to the plasma membrane, where CIBN::pmGFP is localized (C). Panel (D) shows a merge of panels B and C. Scale bars, 10 μm. (E–G) Still frames from a confocal movie showing a top view of the ventral mesoderm of an embryo expressing only CIBN::pmGFP. (E) Toward the end of cellularization, ventral cells at random positions start to show apical constrictions. (F) 8 min later, all ventral cells have dramatically reduced their apical surface. (G) At 13 min, ventral cells invaginate toward the interior of the embryo, creating a fold that is known as ventral furrow. Scale bars, 10 μm. (H–J) Still frames from a confocal movie showing a top view of the ventral mesoderm of an embryo coexpressing mCh::Cry2::5-ptase$_{OCRL}$ and CIBN::pmGFP. Photoactivation is started at the cellularization stage, around 5 min before ventral cells would normally start to constrict. (H), (I), and (J) show the embryo at the onset of photoactivation, at 8 min, and at 22 min after the beginning of photoactivation, respectively. Scale bars, 10 μm.

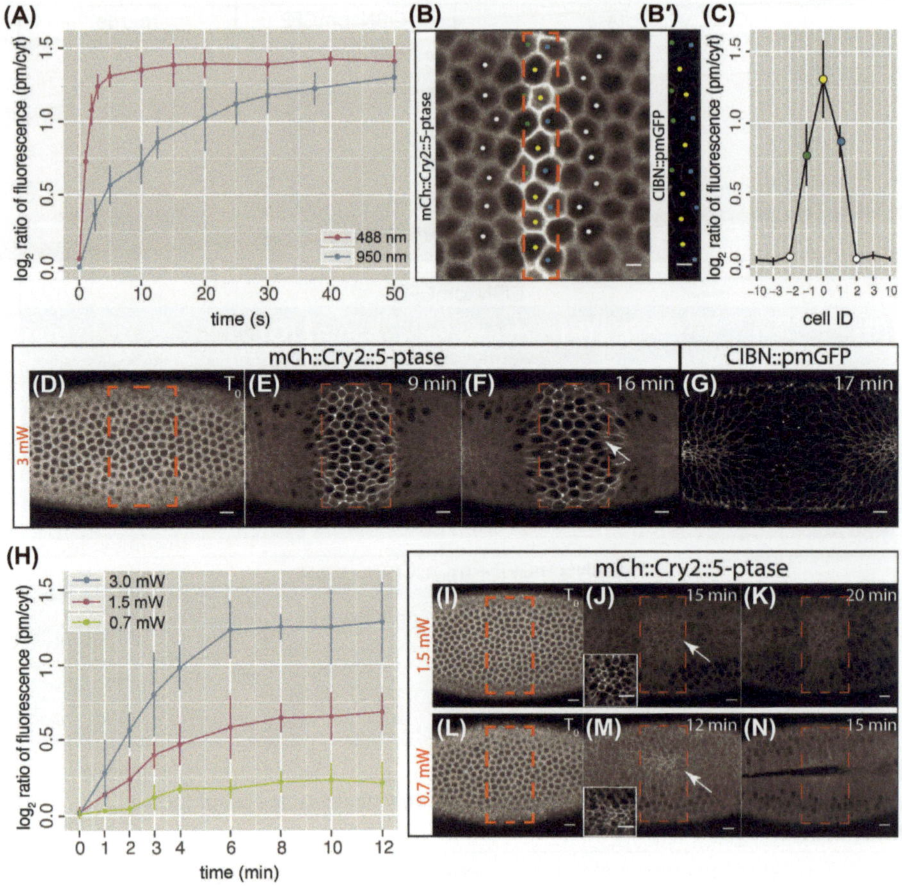

FIGURE 5 Optogenetic activation using two-photon illumination ensures spatially precise and intensity-dependent modulation of apical constriction.

(A) Average photoactivation levels, measured as plasma membrane recruitment of mCh::Cry2::5-ptase$_{OCRL}$, on illumination in one photon (488 nm) or two photon (950 nm). The photoactivation protocols used are described in the main text. X-axis displays the time of illumination, y-axis displays the $\log_2$ ratio of the average plasma membrane versus cytosol intensity of mCh::Cry2::5-ptase$_{OCRL}$. Pooled data are represented as mean $\pm$ SD, $n \geq 3$ embryos for each condition. (B) Top view of an embryo at the blastoderm stage coexpressing mCh::Cry2::5-ptase$_{OCRL}$ and CIBN::pmGFP after 10 min of photoactivation using the two-photon protocol described in the main text. mCh::Cry2::5-ptase$_{OCRL}$ is shown. The *dashed box* represents the photoactivated area. (B′) CIBN::pmGFP is shown. Scale bars, 10 μm. (C) Average photoactivation levels, measured as plasma membrane recruitment of mCh::Cry2::5-ptase$_{OCRL}$, with respect to the photoactivated area in (B). X-axis shows cell IDs. Cell ID = 0 identifies cells that are completely included in the photoactivation area and are marked with a *yellow dot* in (B). Cell IDs = 1 and −1 identify cells that are partially included in the photoactivation area and are marked with a *blue* and a *green dot* in (B), respectively.

absorption of two photons makes two-photon activation less effective than one-photon activation. By adapting the imaging parameters, as described later, we could achieve levels of optogenetic activation comparable to those obtained with one-photon illumination, although with slower kinetics (Fig. 5A).

Cell IDs = 2, 3, 10 and −2, −3, −10 identify cells that are 2, 3, and 10 cells away from the photoactivated area, respectively. Cells with ID = 2 and −2 are marked with a *white dot* in (B). Y-axis shows the $\log_2$ ratio of the average plasma membrane versus cytosol intensity of mCh::Cry2::5-ptase$_{OCRL}$. Pooled data are represented as mean ± SD, $n \geq 3$ embryos for each condition. (D–F) Still frames from a confocal movie showing a top view of the ventral mesoderm of an embryo coexpressing mCh::Cry2::5-ptase$_{OCRL}$ and CIBN::pmGFP. A subgroup of ~80 cells (*dashed box*) within the ventral furrow primordium is photoactivated with the two-photon activation protocol described in the main text at a laser power of 3 mW. (D), (E), and (F) show the embryo before photoactivation, at 9 min, and at 16 min after the beginning of photoactivation, respectively. (G) Confocal image of the same embryo 17 min after the beginning of photoactivation. CIBN::pmGFP is shown. Scale bars, 10 μm. (H) Average photoactivation levels, measured as plasma membrane recruitment of mCh::Cry2::5-ptase$_{OCRL}$, on two-photon illumination at different laser powers (3 mW, 1.5 mW, 0.7 mW). Photoactivation is achieved as described in the main text. X-axis displays time, y-axis displays the $\log_2$ ratio of the average plasma membrane versus cytosol intensity of mCh::Cry2::5-ptase$_{OCRL}$. Pooled data are represented as mean ± SD, $n \geq 3$ embryos for each condition. (I–N) Still frames from a confocal movie showing a top view of the ventral mesoderm of an embryo coexpressing mCh::Cry2::5-ptase$_{OCRL}$ and CIBN::pmGFP. (I–K) A subgroup of ~80 cells (*dashed box*) within the ventral furrow primordium is photoactivated with the two-photon activation protocol described in the main text at a laser power of 1.5 mW. Panels (I), (J), and (K) show the embryo before photoactivation, at 15 min, and at 20 min after the beginning of photoactivation, respectively. (L–N) A subgroup of ~80 cells (*dashed box*) within the ventral furrow primordium is photoactivated with the two-photon activation protocol described in the main text at a laser power of 0.7 mW. Panels (L), (M), and (N) show the embryo before photoactivation, at 12 min, and at 15 min after the beginning of photoactivation, respectively. Insets in panels (J) and (M) display zoom-ins of the photoactivated areas indicated by the *white arrows*. Contrast has been enhanced to better show cell outlines. Embryos activated with lower laser powers display a higher number of constricting cells, as indicated by the *white arrows* (compare panels F, J, and M). Different from panel F, where photoactivated cells fail to constrict, in panel J cells that are photoactivated at suboptimal levels constrict but fail to do so in a coordinated and productive manner, resulting in the blockade of ventral furrow formation. In contrast, cells that are photoactivated at the lowest level (panel M) constrict productively and ventral furrow form with normal kinetics. Scale bars, 10 μm.

3b. Photoactivation with two-photon illumination can be achieved using light at $\lambda = 900$ and 950 nm. 950 nm has proven to be the most effective wavelength for triggering the plasma membrane translocation of mCh::Cry2::5-ptase$_{OCRL}$. Optimal two-photon activation is achieved by illuminating five consecutive z-stacks separated by 1 μm (the first z-stack being 5 μm from the apical surface of the embryo), using a femtosecond pulsed laser at a repetition rate of 80 MHz and $\lambda = 950$ nm (laser power: 3 mW, measured 1 cm from the objective). Each z-stack is illuminated at a scanning speed of 1×10^{-6} s/pixel and averaging of 2 in bidirectional mode, which corresponds to a scan time of 500 ms per z-stack (2.5 s in total for five z-stacks). Two consecutive photoactivations are intervalled with a cycle of imaging of mCh::Cry2::5-ptase$_{OCRL}$ at 561 nm, followed by a 30 s dark interval.
This two-photon illumination protocol ensures that photoactivation remains limited to the illuminated area, as shown in Fig. 5B and C. This feature makes it possible to trigger the plasma membrane translocation of mCh::Cry2::5-ptase$_{OCRL}$ in a specific group of cells within the ventral furrow primordium, and block apical constriction specifically in those cells (Fig. 5D–F). At the end of the experiment, cell outlines in nonactivated areas can be inspected by taking a snapshot of the entire embryo with a 488 nm laser (Fig. 5G).

By employing this protocol, it is also possible to modulate the extent to which apical constriction is inhibited by simply varying the power of the light input. Indeed, decreasing laser powers (3 mW, 1.5 mW, and 0.7 mW) result in decreasing levels of photoactivation, measured as the $\log_2$ ratio of the average plasma membrane versus cytosol intensity of mCh::Cry2::5-ptase$_{OCRL}$ (Fig. 5H). Embryos displaying lower levels of photoactivation show less effective inhibition of apical constrictions compared to embryos that are photoactivated at optimal levels (compare photoactivated cells in Fig. 5 panels D–F with those in Fig. 5 panels I–N).

In summary, we described two protocols for optogenetic inhibition of apical constriction that are based on either one- or two-photon illumination. Photoactivation in one-photon enables rapid (seconds timescale) plasma membrane recruitment of mCh::Cry2::5-ptase$_{OCRL}$ in the entire embryo and blockade of ventral furrow formation. Two-photon illumination allows photoactivation and inhibition of apical constriction with high spatial precision (single cell resolution) and on the subminute timescale. Moreover, as longer wavelengths can penetrate deeper into tissue than shorter ones, two-photon illumination allows, in principle, photoactivation of cells or subcellular structures located deeper inside the embryo. Lastly, by increasing or decreasing the laser power, it is possible to interfere gradually with apical constriction and therefore test the effect of fine perturbations of apical constrictions on morphogenetic responses.

1.4.1 Materials

- Confocal laser-scanning microscope (LSM 780 NLO, Carl Zeiss), equipped with:
    - Argon multiline 458/488/514 nm laser
    - HeNe 561 nm laser

  - Multiphoton femtosecond (140 fs) pulsed laser, 80 MHz repetition rate (Chamaleon, Coherent Inc.)
- C-Apochromat 63x/NA 1.2 water immersion objective (Carl Zeiss)
- Deep Amber lighting filter 104 (Cabledelight Ltd., available on Amazon.com)
- Zen 2011 software for image acquisition (Carl Zeiss)
- Laser power meter (Laser Check, Edmund Optics Inc.—catalog no. 54018)

CONCLUSION

We presented a step-by-step protocol for optogenetic inhibition of apical constriction during *Drosophila* embryogenesis. Light-mediated plasma membrane recruitment of the catalytic domain of the 5-ptase OCRL enables the perturbation of actin-dependent morphogenetic processes with both spatial and temporal precision. Indeed, by applying this method to locally modulate apical constriction during gastrulation, we were able to address the role of geometrical constraints and cell–cell interaction during ventral furrow formation (Guglielmi et al., 2015). Given that apical constriction, and more in general actin dynamics, are paramount for many other morphogenetic processes (reviewed in Sawyer et al., 2010), we expect that this method will help addressing the regulation of tissue dynamics in other contexts as well. For example, during *Drosophila* dorsal closure, which is a model for the study of wound healing, reduction of the apical surface in amnioserosa cells has been shown to facilitate the closing of the dorsal gap. Indeed, pulses of apical constriction correlate with a decrease in total area of the amnioserosa tissue (David, Wang, Feng, & Harris, 2013; Gorfinkiel, Blanchard, Adams, & Martinez Arias, 2009; Mateus, Gorfinkiel, Schamberg, & Martinez Arias, 2011), and slower rates of closure have been observed on laser ablation of multiple amnioserosa cells (Hutson et al., 2003). However, it is not clear how forces generated within the amnioserosa are transmitted along the tissue and to lateral ectodermal cells, which are in direct contact with the amnioserosa. Optogenetic modulation of apical constriction in a subgroup of amnioserosa cells would provide a powerful approach to perturb the mechanical equilibrium, and probe force transmission within the amnioserosa and between the amnioserosa and the adjacent ectodermal tissue. Computer simulations coupled with laser ablation experiments showed that tissue tension impacts on individual amnioserosa cell behavior, most likely through a yet unexplored mechanosensitive mechanism (Solon et al., 2009). Mechanosensation has been shown in both cell culture (Dupont et al., 2011) and vertebrate developmental systems (Porazinski et al., 2015), where it appears to be controlled by YAP/TAZ. However, it is unknown whether this is a general principle through which tissues and organs are shaped or if other molecular mechanisms are involved.

The modulation of apical constriction at different laser powers offered by this optogenetic approach represents a powerful means to reduce tissue tension at will in single cells or collectives, and monitor the effects on neighbors' behavior in real time. As apical constriction is an evolutionary conserved process from fly gastrulation to vertebrate neurulation (Baker & Schroeder, 1967; Moore, Stanisstreet, & Evans, 1987; Schoenwolf & Franks, 1984), the application of this optogenetic approach to other organisms should facilitate the study of dynamics and propagation of forces during animal morphogenesis.

Finally, by combining precise optogenetic perturbations, such as the one described in this chapter, with multiview SPIM (Krzic, Gunther, Saunders, Streichan, & Hufnagel, 2012; Rauzi et al., 2015), it will be possible to study cell and tissue interactions at the embryo-scale level. The integration of such data with in silico modeling will help us gain a more detailed understanding of the mechanisms and mechanics-driving tissue remodeling during animal morphogenesis.

ACKNOWLEDGMENTS

We acknowledge all the members of the De Renzis laboratory for discussion and E. Loeser and T. Quinkler for technical assistance. We acknowledge the Advanced Light Microscopy Facility and the Developmental Biology Unit at the European Molecular Biology Laboratory for support. We thank E. Wieschaus for the maternal tubulin Gal4 stock and the Drosophila Genomics Resource Center for providing cDNAs. This work was supported by a Human Frontier Science Program Career Development Award to SDR.

REFERENCES

Baker, P. C., & Schroeder, T. E. (1967). Cytoplasmic filaments and morphogenetic movement in the amphibian neural tube. *Developmental Biology, 15*(5), 432–450.

Barone, V., & Heisenberg, C. P. (2012). Cell adhesion in embryo morphogenesis. *Current Opinion in Cell Biology, 24*(1), 148–153. http://dx.doi.org/10.1016/j.ceb.2011.11.006.

Cai, D., Chen, S. C., Prasad, M., He, L., Wang, X., Choesmel-Cadamuro, V., … Montell, D. J. (2014). Mechanical feedback through E-cadherin promotes direction sensing during collective cell migration. *Cell, 157*(5), 1146–1159. http://dx.doi.org/10.1016/j.cell.2014.03.045.

Campos-Ortega, J. A., & Hartenstein, V. (1997). *The embryonic development of* Drosophila melanogaster (2nd ed.).

Colombelli, J., & Solon, J. (2013). Force communication in multicellular tissues addressed by laser nanosurgery. *Cell and Tissue Research, 352*(1), 133–147. http://dx.doi.org/10.1007/s00441-012-1445-1.

David, D. J., Wang, Q., Feng, J. J., & Harris, T. J. (2013). Bazooka inhibits aPKC to limit antagonism of actomyosin networks during amnioserosa apical constriction. *Development, 140*(23), 4719–4729. http://dx.doi.org/10.1242/dev.098491.

Dupont, S., Morsut, L., Aragona, M., Enzo, E., Giulitti, S., Cordenonsi, M., … Piccolo, S. (2011). Role of YAP/TAZ in mechanotransduction. *Nature, 474*(7350), 179–183. http://dx.doi.org/10.1038/nature10137.

Gorfinkiel, N., Blanchard, G. B., Adams, R. J., & Martinez Arias, A. (2009). Mechanical control of global cell behaviour during dorsal closure in Drosophila. *Development, 136*(11), 1889–1898. http://dx.doi.org/10.1242/dev.030866.

Guglielmi, G., Barry, J. D., Huber, W., & De Renzis, S. (2015). An optogenetic method to modulate cell contractility during tissue morphogenesis. *Developmental Cell*. http://dx.doi.org/10.1016/j.devcel.2015.10.020.

Hutson, M. S., Tokutake, Y., Chang, M. S., Bloor, J. W., Venakides, S., Kiehart, D. P., & Edwards, G. S. (2003). Forces for morphogenesis investigated with laser microsurgery and quantitative modeling. *Science, 300*(5616), 145–149. http://dx.doi.org/10.1126/science.1079552.

Idevall-Hagren, O., Dickson, E. J., Hille, B., Toomre, D. K., & De Camilli, P. (2012). Optogenetic control of phosphoinositide metabolism. *Proceedings of National Academy of Sciences of the United States of America, 109*(35), E2316–E2323. http://dx.doi.org/10.1073/pnas.1211305109.

Kennedy, M. J., Hughes, R. M., Peteya, L. A., Schwartz, J. W., Ehlers, M. D., & Tucker, C. L. (2010). Rapid blue-light-mediated induction of protein interactions in living cells. *Nature Methods, 7*(12), 973–975. http://dx.doi.org/10.1038/nmeth.1524.

Krzic, U., Gunther, S., Saunders, T. E., Streichan, S. J., & Hufnagel, L. (2012). Multiview light-sheet microscope for rapid in toto imaging. *Nature Methods, 9*(7), 730–733. http://dx.doi.org/10.1038/nmeth.2064.

Lecuit, T., Lenne, P. F., & Munro, E. (2011). Force generation, transmission, and integration during cell and tissue morphogenesis. *Annual Review of Cell and Developmental Biology, 27*, 157–184. http://dx.doi.org/10.1146/annurev-cellbio-100109-104027.

Leptin, M., & Grunewald, B. (1990). Cell shape changes during gastrulation in Drosophila. *Development, 110*(1), 73–84.

Martin, A. C., Gelbart, M., Fernandez-Gonzalez, R., Kaschube, M., & Wieschaus, E. F. (2010). Integration of contractile forces during tissue invagination. *Journal of Cell Biology, 188*(5), 735–749. http://dx.doi.org/10.1083/jcb.200910099.

Martin, A. C., Kaschube, M., & Wieschaus, E. F. (2009). Pulsed contractions of an actin-myosin network drive apical constriction. *Nature, 457*(7228), 495–499. http://dx.doi.org/10.1038/nature07522.

Mason, F. M., Tworoger, M., & Martin, A. C. (2013). Apical domain polarization localizes actin-myosin activity to drive ratchet-like apical constriction. *Nature Cell Biology, 15*(8), 926–936. http://dx.doi.org/10.1038/ncb2796.

Mateus, A. M., Gorfinkiel, N., Schamberg, S., & Martinez Arias, A. (2011). Endocytic and recycling endosomes modulate cell shape changes and tissue behaviour during morphogenesis in Drosophila. *PLoS One, 6*(4), e18729. http://dx.doi.org/10.1371/journal.pone.0018729.

Moore, D. C., Stanisstreet, M., & Evans, G. E. (1987). Morphometric analyses of changes in cell shape in the neuroepithelium of mammalian embryos. *Journal of Anatomy, 155*, 87–99.

Munjal, A., & Lecuit, T. (2014). Actomyosin networks and tissue morphogenesis. *Development, 141*(9), 1789–1793. http://dx.doi.org/10.1242/dev.091645.

Porazinski, S., Wang, H., Asaoka, Y., Behrndt, M., Miyamoto, T., Morita, H., ... Furutani-Seiki, M. (2015). YAP is essential for tissue tension to ensure vertebrate 3D body shape. *Nature, 521*(7551), 217–221. http://dx.doi.org/10.1038/nature14215.

Rauzi, M., Krzic, U., Saunders, T. E., Krajnc, M., Ziherl, P., Hufnagel, L., & Leptin, M. (2015). Embryo-scale tissue mechanics during Drosophila gastrulation movements. *Nature Communications, 6*, 8677. http://dx.doi.org/10.1038/ncomms9677.

Rauzi, M., Verant, P., Lecuit, T., & Lenne, P. F. (2008). Nature and anisotropy of cortical forces orienting Drosophila tissue morphogenesis. *Nature Cell Biology, 10*(12), 1401–1410. http://dx.doi.org/10.1038/ncb1798.

Reversi, A., Loeser, E., Subramanian, D., Schultz, C., & De Renzis, S. (2014). Plasma membrane phosphoinositide balance regulates cell shape during Drosophila embryo morphogenesis. *Journal of Cell Biology, 205*(3), 395–408. http://dx.doi.org/10.1083/jcb.201309079.

Rorth, P. (1998). Gal4 in the Drosophila female germline. *Mechanisms of Development, 78*(1–2), 113–118.

Saarikangas, J., Zhao, H., & Lappalainen, P. (2010). Regulation of the actin cytoskeleton-plasma membrane interplay by phosphoinositides. *Physiological Reviews, 90*(1), 259–289. http://dx.doi.org/10.1152/physrev.00036.2009.

Sawyer, J. M., Harrell, J. R., Shemer, G., Sullivan-Brown, J., Roh-Johnson, M., & Goldstein, B. (2010). Apical constriction: a cell shape change that can drive morphogenesis. *Developmental Biology, 341*(1), 5–19. http://dx.doi.org/10.1016/j.ydbio.2009.09.009.

Schoenwolf, G. C., & Franks, M. V. (1984). Quantitative analyses of changes in cell shapes during bending of the avian neural plate. *Developmental Biology, 105*(2), 257–272.

Solon, J., Kaya-Copur, A., Colombelli, J., & Brunner, D. (2009). Pulsed forces timed by a ratchet-like mechanism drive directed tissue movement during dorsal closure. *Cell, 137*(7), 1331–1342. http://dx.doi.org/10.1016/j.cell.2009.03.050.

Spahn, P., & Reuter, R. (2013). A vertex model of Drosophila ventral furrow formation. *PLoS One, 8*(9), e75051. http://dx.doi.org/10.1371/journal.pone.0075051.

Sweeton, D., Parks, S., Costa, M., & Wieschaus, E. (1991). Gastrulation in Drosophila: the formation of the ventral furrow and posterior midgut invaginations. *Development, 112*(3), 775–789.

Toettcher, J. E., Voigt, C. A., Weiner, O. D., & Lim, W. A. (2011). The promise of optogenetics in cell biology: interrogating molecular circuits in space and time. *Nature Methods, 8*(1), 35–38. http://dx.doi.org/10.1038/nmeth.f.326.

Toettcher, J. E., Weiner, O. D., & Lim, W. A. (2013). Using optogenetics to interrogate the dynamic control of signal transmission by the Ras/Erk module. *Cell, 155*(6), 1422–1434. http://dx.doi.org/10.1016/j.cell.2013.11.004.

Wang, X., He, L., Wu, Y. I., Hahn, K. M., & Montell, D. J. (2010). Light-mediated activation reveals a key role for Rac in collective guidance of cell movement in vivo. *Nature Cell Biology, 12*(6), 591–597. http://dx.doi.org/10.1038/ncb2061.

CHAPTER

Micropipette aspiration: a unique tool for exploring cell and tissue mechanics in vivo

10

K. Guevorkian*, J.-L. Maître[§,1]

*Université de Strasbourg, Centre National de la Recherche Scientifique, UMR 7104, Institut National de la Santé et de la Recherche Médicale, U964, Illkirch, France

[§]Institut Curie, PSL Research University, CNRS, UMR 3215, INSERM, U934, Paris, France

[1]Corresponding author: E-mail: jean-leon.maitre@curie.fr

CHAPTER OUTLINE

Methods in Cell Biology, Volume 139, ISSN 0091-679X, http://dx.doi.org/10.1016/bs.mcb.2016.11.012

Abstract

Cell and tissue mechanical properties are paramount in controlling morphogenesis. Microaspiration techniques allow measuring the absolute values of mechanical properties in space and time in vivo. Here, we explain how to build a microaspiration setup that can be used for both cellular and tissue scale measurements. At the cellular scale, microaspiration allows the mapping in space and time of surface tensions of individual interfaces within a tissue to understand the forces shaping it. At the tissue scale, microaspiration can be used to measure macroscopic mechanical properties such as the viscoelasticity and tissue surface tension that regulate the dynamics of tissue deformation. Based on a simple and cost-effective apparatus, these two complementary microaspiration techniques provide unique tools for exploring cell and tissue mechanics in vivo.

INTRODUCTION

Cell and tissue mechanical properties control their shape, behavior, and function (Gonzalez-Rodriguez, Guevorkian, Douezan, & Brochard-Wyart, 2012; Heisenberg & Bellaïche, 2013). During embryonic morphogenesis, changes in mechanical properties can elongate (Collinet, Rauzi, Lenne, & Lecuit, 2015), fold (Rauzi et al., 2015), stretch (Behrndt et al., 2012), round up (Maître, Niwayama, Turlier, Nédélec, & Hiiragi, 2015), segregate (Maître et al., 2016), or mix tissues (Levayer, Hauert, & Moreno, 2015). Tissue mechanics can also be read as patterning cues and directly control cell differentiation (Swift et al., 2013) and proliferation (Wolfenson et al., 2015). In addition, changes in cell and tissue mechanics often occur in pathologies such as cancer (Guck et al., 2005). Therefore, being able to measure quantitatively cell and tissue mechanical properties is essential to understand a wide variety of biological phenomena.

To measure the mechanical properties of a material, one needs to deform (strain) a sample with a known force (stress), which can be either compressive or tensile (Sugimura, Lenne, & Graner, 2016). In 1954, Mitchison and Swann described the "cell elastimeter," a microaspiration device, which they used to pull on the cell surface with a controlled pressure. Using the cell elastimeter, Mitchison and Swann (1955) could measure quantitatively the mechanical properties of a sea urchin embryo from fertilization to the first cleavage, which helped better understanding the forces driving cytokinesis. Since then, microaspiration setups analogous to the cell elastimeter have been used to measure the mechanical properties not only of a variety of isolated cells but also of cells within tissues. Microaspiration setups were used to measure cell surface tensions in the pN/μm range for granulocytes (Evans & Yeung, 1989) to the nN/μm range for oocytes (Larson et al., 2010). Microaspiration has also been used to evaluate the elasticity of embryonic tissues, ranging from tens to hundreds of Pa, for example, during *Xenopus laevis* gastrulation (von Dassow & Davidson, 2009) or in developing chicken heart and brain (Majkut et al., 2013). Microaspiration can also be used to determine the apparent viscosity of isolated cells, with granulocytes at a few kPa.s (Evans & Yeung, 1989), and whole tissues, with aggregates of sarcoma cells at hundreds of kPa.s (Guevorkian, Colbert, Durth, Dufour, & Brochard-Wyart, 2010). Microaspiration is a nondestructive method applicable to cells, dissected tissues, or embryos. Therefore, the measurements can be repeated on the same sample to track its mechanical properties over time (Maître et al., 2015). Last but not least, microaspiration setups are simple to build and use, cost-effective and can be combined with most inverted microscopes to study the shape changes of the sample and relevant molecules while measuring cell and tissue mechanics (Maître et al., 2016). Altogether, microaspiration is a versatile tool to probe cells and tissues over a broad range of their mechanical properties in vivo.

In this chapter, we will describe a protocol for building a microaspiration setup and how to use it to map the tensions of individual interfaces within a tissue or to measure the bulk mechanical properties of a tissue.

1. BUILDING AND PREPARING A MICROASPIRATION SETUP

Microaspiration setups have been built for over 60 years (Mitchison & Swann, 1954) and excellent protocols, for example, describing a microaspiration setup based on a syphon mechanism and aimed at measuring the surface tension of isolated cells (Kee & Robinson, 2013), are already available. The microaspiration setup described here includes a microfluidic pump, in addition to a syphon, similarly to a previously described dual microaspiration setup (Biro & Maître, 2015). This allows for faster and easier pressure control, improved pressure stability, and opens up the possibility for some automation (Biro & Maître, 2015) while keeping the total price of the microaspiration setup itself (omitting the microscope and pipette preparation machines) around 5000€.

1.1 PREPARING MICROPIPETTES

For microaspiration of cells and tissues, a glass micropipette is prepared in three steps: first, a glass needle is pulled; second, the needle is cut and bent; and last, if needed, the micropipette's surface is passivated.

1.1.1 Materials

- Glass capillaries (without filament, for example, World Precision Instruments TW100-3)
- Micropipette-flushing needle (for example, MicroFil MF28G-5 from World Precision Instruments)
- Optional: coating solution [for example, fetal bovine serum (FBS), bovine serum albumin or polyethylene glycol (PLL-PEG from SuSoS, Switzerland)]
- Cleaning solution (water, PBS, or cell culture medium)

1.1.2 Equipment

- Pipette puller (for example, P-97 Flaming Brown from Sutter Instrument Co.)
- Microforge (for example, MF-900 from Narishige or MF-5 from Microdata Instrument)
- Pipette storage container, required to keep micropipettes clean from dust or immersed in cleaning solution (commercial units can be obtained from World Precision Instruments; alternatively, they can easily be custom-made using a standard 50 mL tube, where adhesive clay is deposited on the inside of the lid).
- Optional: pipette-coating container (can be custom-made as above with additional adhesive clay molded inside the tube to hold a round-bottom 2 mL tube in place, containing a few hundred μL of the coating solution).

1.1.3 Methods

1. Needles are pulled from glass capillaries using the pipette puller so that a centimeter long taper is obtained (for example, using the following program on the P-97 Flaming Brown from Sutter Instrument Co. for 0.75 mm ID and 1 *mm* OD capillaries: Ramp +3−5, Pull 50, Velocity 50, Time 50, and Pressure 500).
2. Depending on the desired pipette diameter one of the following techniques can be used to cut the pipette tip.
   - For small diameters, ranging between 1 and 50 μm, suitable for experiments with individual cells and small tissues, we use a microforge to cut the taper of the needle at the required micropipette radius. A ~0.5 mm glass bead sitting on the microforge filament is positioned next to the needle at the required breaking point at a distance of ~20 μm with an approaching angle of ~20 degrees. Thereafter, the filament and bead are heated up for less than 1 s so that the glass bead is slightly displaced due to thermal expansion and briefly contacts the needle. After contact and extinction of the heat source, the retraction of the bead causes a blunt break at the contact point on the needle.
   - For pipette diameters ranging between 50 and 100 μm suitable for tissue manipulation, a ceramic tile (Sutter Instrument) or a second pipette can be used to break the tip at the desired size. Score the taper of the pipette at the desired cut diameter before bending the tip until it breaks. The pipette tip can be fire polished by approaching the tip of the pipette to the heated wire of a microforge.
3. Using the microforge, the micropipette tip is bent such that, when mounted on the micromanipulator, the tip is parallel to the bottom of the dish (typically 20−40 degrees). The microforge filament is placed at a 90 degree angle and at ~100 μm from the tip of the micropipette. The filament of the microforge is heated up and progressively brought closer to the micropipette until it bends to the required angle.

4. Optional: micropipettes can be passivated to reduce adhesion to the glass. To this end, the coating substance, concentration, and duration must be chosen specifically for each sample. Stick the micropipette on the lid of a pipette-coating container and dip it in the coating solution. For example, efficient passivation with FBS at RT takes ~5 min. Alternatively, for PLL–PEG coating, use 0.1 mg/mL PLL-PEG diluted in 10 μM HEPES at pH 7.5 at 4°C for 1 h on plasma cleaned or freshly pulled pipettes. To avoid clotting of the micropipette or contamination of the experiment with coating solution, the micropipettes are rinsed with cleaning solution (ideally the solution that is used for filling the aspiration setup, such as PBS).

1.2 ASSEMBLING THE ASPIRATION SETUP

Fig. 1 shows the schematic diagram of the microaspiration setup. The microaspiration technique relies on the accurate calibration of the aspiration pressure. The micropipette needs to be carefully installed and hermetically connected to the tubing before the intermediate reservoir and microfluidic pump are adjusted to calibrate the aspiration pressure. The calibration should allow flexibility in the applied pressures,

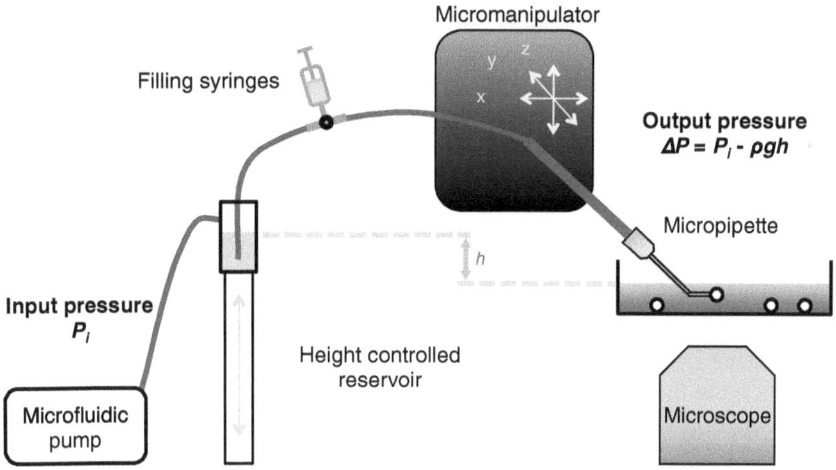

FIGURE 1 Microaspiration setup.

Schematic diagram of the microaspiration setup showing the micromanipulator mounted on the microscope. The micropipette is connected to a liquid reservoir, which is mounted on a stage that can be translated vertically with micrometer precision. The elevation h of the reservoir as compared to the culture dish containing the sample acts as a siphon and provides a positive pressure $P_{res} = \rho g h$, where ρ is the liquid volumetric mass density and g is Earth's gravity pull. The liquid reservoir is connected to a microfluidic pump, which provides a negative pressure P_i. The resulting pressure at the tip of the micropipette is $\Delta P = P_i + \rho g h$.

both in the positive and negative ranges, and a null differential pressure should be clearly defined and accurately attained.

1.2.1 Materials

- PBS/medium
- Micropipette-flushing needle (for example, MicroFil MF28G-5 from World Precision Instruments)

1.2.2 Equipment

- Inverted microscope that is capable of transmitted light imaging and with appropriate incubation conditions
- Micromanipulator allowing fine movements in three dimensions (for example, the Narishige MMO-202ND)
- Microfluidic pump able to provide at least 5 kPa with a resolution of a few Pa (for example, the Fluigent MFCS)
- Pipette holder (for example, Eppendorf capillary holder with associated grip head is compatible with male M6 luers)
- Tubing and luers to connect the pipette holders to a three-way valve (M6 luers and FEP tubes of 1.3 OD are compatible with above example equipment)
- Syringe to connect to the three-way valve to fill the tubing and reservoir
- Intermediate reservoir (for example, a 15 mL tube) of which the height is controlled with a linear translation stage with μm resolution (for example, a 50 mm travel stage and associated microscale acquired from Newport and mounted on a fixed stand or the environmental chamber of the microscope)

1.3.3 Methods

1. Position the three-way valve to connect the syringe and intermediate reservoir, and fill it up with PBS/medium carefully so that no air bubbles are introduced.
2. Switch the three-way valve to connect the syringe and micropipette tubing, and fill it up with PBS/medium carefully so that no air bubbles are introduced.
3. Fill the micropipette with PBS/medium using the micropipette-flushing needle.
4. Mount the micropipette onto the pipette holder.
5. Connect the pipette holder to the tubing carefully so that no air bubbles are introduced.
6. Bring the micropipette to a safe distance above the sample.
7. Power the microfluidic pump on.
8. Connect the intermediate reservoir to the microfluidic pump.
9. Adjust the height of the intermediate reservoir so that the micropipette expels liquid without any pressure being provided by the microfluidic pump.

10. Adjust the pressure of the microfluidic pump until zero flux is observed in the micropipette to set P_0, the zero flux pressure. To this end, small beads (<1 μm) or particles/debris floating in the dish can initially be aspirated and the pressure adjusted until no movement is observable. During this calibration process, care for the beads or particles not to stick to the walls of the micropipette. Finding and setting P_0 can be difficult as weak residual eddies can remain even in the absence of bubbles or other perturbations in the system. One can consider that the zero flux pressure has been attained when only a few Pa (smaller than the pressure steps used during experiments) are sufficient to invert these weak residual flows. The resulting error on the measurements will then be negligible (see Discussion).

2. SUBCELLULAR SURFACE TENSION MAPPING

During tissue morphogenesis, cellular interfaces change their tensions, which causes tissues to change shape. Microaspiration allows measuring the effective surface tension of cells by deforming their surface with a known pressure.

2.1 THEORY

Cells and tissues can show liquid-like behaviors and therefore an effective surface tension can be characterized. Using microaspiration, the surface tension at the cell—medium interface γ_{cm} is determined from the following force balance equation: $\gamma_{cm} = P_c/2(1/R_p - 1/R_c)$, where R_c is the resting radius of curvature of the interface in contact with micropipette, P_c is the critical pressure at which the cell deformation reaches R_p, the micropipette radius. This was performed for a variety of cells in suspension (Evans & Yeung, 1989; Dai, Ting-Beall, Hochmuth, Sheetz, & Titus, 1999; Larson et al., 2010; Tinevez et al., 2009) or within a tissue (Maître et al., 2015, 2016). In the case of tissues, the surface tension of multiple contacting cells can be measured. This allows calculating the tension γ_{cc} at the cell—cell interface of cells previously measured by microaspiration, or more generally to propagate from interface to interface as long as the tension of at least two interfaces γ_{cm1} and γ_{cm2} are known. To do so, one can apply the following force balance equation: $\gamma_{cc} = -\gamma_{cm1} \cos(\theta_2) - \gamma_{cm2} \cos(\theta_1)$, where θ_1 and θ_2 are the angles at the contact line between interfaces of tension γ_{cm1} and γ_{cc}, and γ_{cm2} and γ_{cc}, respectively, as shown in Fig. 2. Using this method, charting the tensions of all interfaces of a group of cells becomes possible. Moreover, since micropipette aspiration is not damaging the sample, it can be repeated to build a complete spatio-temporal map of surface tensions. This can be used to determine the relative contribution of distinct interfaces to a shape change (Maître et al., 2015) or follow the fate of cells after having characterized their respective surface tensions (Maître et al., 2016). The main assumption of this method is the liquid-like behavior of cells. To verify this, one can apply a pressure higher than the critical pressure P_c. This will cause the cell to

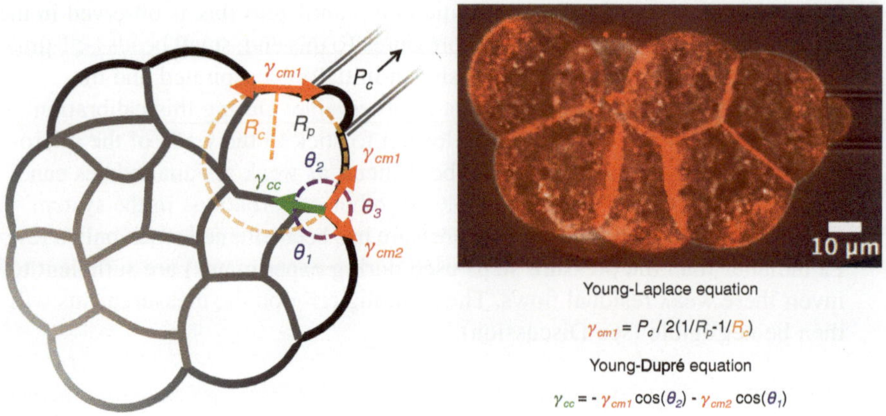

FIGURE 2 Surface tension mapping.

Schematic diagram of the surface tension mapping technique. Using a micropipette of radius R_p, the surface of cell of radius R_c is bent to a deformation of size R_p at the critical pressure P_c. The Young–Laplace equation gives the surface tension at the cell–medium interface $\gamma_{cm} = P_c/2(1/R_p - 1/R_c)$. After measuring two contacting cells of the aggregate, the tension at their cell–cell contact can be calculated following the Young–Dupré equation: $\gamma_{cc} = -\gamma_{cm1} \cos(\theta_2) - \gamma_{cm2} \cos(\theta_1)$, where θ_2 and θ_1 are the internal contact angles of the cells of surface tension γ_{cm1} and γ_{cm2}, respectively. θ_3 is the external contact angle. Picture of an 8-cell stage mouse embryo expressing a membrane marker (red). The right-most blastomere is held by a micropipette aspirating with P_c during a surface tension mapping experiment (Movie 1).

flow inside the micropipette at constant velocity if the cell behaves like a liquid (this can be used to measure additional mechanical properties, see next section).

2.2 MATERIALS

- Cell aggregates in suspension
- Cell culture medium
- Mineral oil for prolonged experiments
- Nonadhesive glass bottom culture dish
- Micropipette typically between one-third and one-eighth of the radius of a cell

2.3 METHODS

1. Measure the radius of the micropipette opening R_p by drawing a parallel line between the inner walls close to the tip [for example, using Fiji's line tool (Schindelin et al., 2012)].

2. Using the micropipette, orient the aggregate so that one cell is visible and accessible to the micropipette.
3. Take a picture of the cell to be measured, and measure the radius of curvature of the cell R_c by fitting a circle on the cell–medium interface (for example, using Fiji's circle tool), the angles θ at the contact lines that are perpendicular to the imaging plane (for example, using Fiji's angle tool).
4. Approach the cell with a low suction pressure (10–30 Pa).
5. Increase the pressure stepwise until the aspirated part of the cell corresponds to the radius of the micropipette (it should form a hemisphere). Note this pressure as the critical pressure P_c. Due to cells viscosity, pressure changes do not translate instantaneously to the final deformation. Therefore, one should wait a few tens of seconds between pressure changes for the deformation to equilibrate.
6. After releasing the cell, watch that there is no adhesion with the micropipette and that the cell recovers its spherical surface (typically within a few seconds to minutes).
7. Calculate the surface tension at the cell–medium interface of the cell using the Young–Laplace equation: $\gamma_{cm} = P_c/2(1/R_p - 1/R_c)$.
8. Repeat the procedure with as many neighboring cells as needed and as long as no shape changes occur.
9. Calculate the tensions of interfaces for which the tensions of two contacting interfaces are known using the following formula: $\gamma_{cc} = -\gamma_{cm1} \cos(\theta_2) - \gamma_{cm2} \cos(\theta_1)$.
10. Repeat the procedure at a later time, for example, after significant shape changes have occurred.

2.4 DISCUSSION

Surface tension mapping is made possible by the unique set of properties of microaspiration (Sugimura et al., 2016). Indeed, unlike laser ablation techniques, microaspiration allows measuring surface tensions in a nondestructive manner and thereby permits tracking cell mechanics over time. Also, unlike atomic force microscope techniques, microaspiration enables the experimenter to easily move and orient freely floating samples to probe different cells from the tissue. Finally, this method reports the absolute values of surface tensions, which is essential to compare different tissues in space and time, and greatly facilitates elaborating theoretical models of tissue shape changes.

During microaspiration, strong pressures can be applied to the cell surface, which can sometimes detach the plasma membrane from the cell actomyosin cortex and induce the formation of a bleb. The formation of blebs can easily be distinguished from the simple deformation of the cell surface due to the applied pressure; a bleb typically grows much larger and faster than a controlled deformation, can grow asymmetrically, typically retracts after about a minute and contains no actomyosin cortex during its growth phase (Charras & Paluch, 2008). Blebs tend to occur more frequently with decreasing R_p, and increasing the micropipette size can solve this problem. However, increasing the micropipette size decreases the resolution of the measurement with identical pressure steps and can make some cells inaccessible to measurement due to the topology of the tissue.

The measurement of a cell surface tension by microaspiration can take a few minutes and therefore the measurement of several cells can take a significant amount of time during which cells may move, deform, or divide. The surface tension mapping method described above relies on steady state throughout the measurements. This method is therefore not suited for rapidly changing tissues for which only a couple of measurements will be possible before significant shape changes occur. However, it is not necessary to measure all of the cells from a cell aggregate to map its surface tension. In fact, three to four measurements and a high-resolution three-dimensional image are in principle enough to infer the tensions of all interfaces by propagating from two known interfaces to an uncharacterized neighboring interface. In this case, one should watch for the measurement errors on determining the surface tension (which depend on the accuracy on finding the critical pressure and the micropipette size) and geometrical parameter (which depends on the imaging conditions), which will also propagate. An essential control will be to calculate the surface tension of cells both by direct microaspiration and by inferring from tensions of neighboring interfaces.

3. MECHANICAL PROPERTIES OF TISSUES

During tissue morphogenesis, cellular contacts are remodeled in response to internal or external stresses, for example, due to changes in interfacial tensions. The dynamics of tissue shape changes depend on its viscoelastic properties, or rheology (Gonzalez-Rodriguez et al., 2012). For example, a tissue with high viscosity will deform slower than a tissue with low viscosity when subjected to a given stress. Therefore, determining the viscoelastic properties of living tissues will help understanding the timescale at which morphogenetic changes occur. Here, we describe how to measure the viscoelastic properties, as well as the tissue surface tension, of a spherical cell aggregate using microaspiration. It is worth noting that this technique can be used on nonspherical tissues as well.

3.1 THEORY

As for single cell microaspiration, when a pressure, ΔP, greater than the critical pressure, P_c, is applied, tissues behaving like a viscoelastic fluid will flow inside a micropipette (Guevorkian et al., 2010). At a given pressure, the rate of aspiration will depend on the tissue resistance to deformation and friction on the pipette wall. Because the pipettes are made nonadhesive, the wall friction can be neglected and only the tissue mechanical properties remain. Typically, after an initial elasticity-dominated response, the tissue flows at constant velocity inside the micropipette (Fig. 3). At that point, the flow rate, $\dot{L}_{asp}$, depends on the viscosity, η, as $\dot{L}_{asp} = \frac{R_p}{3\pi\eta}(\Delta P - P_c)$. If the pressure is set to zero, the tissue will retract under the action of its surface tension, γ, at a velocity $\dot{L}_{ret} = \frac{R_p}{3\pi\eta}P_c$, once again, after an elastic dissipation. From the aspiration $\dot{L}_{asp}$ and retraction $\dot{L}_{ret}$ rates, the viscosity can be calculated as $\eta = \frac{R_p \Delta P}{3\pi\,(\dot{L}_{asp} + \dot{L}_{ret})}$ (Guevorkian et al., 2010). The surface tension of a tissue is measured by evaluating P_c, which, as previously, can be

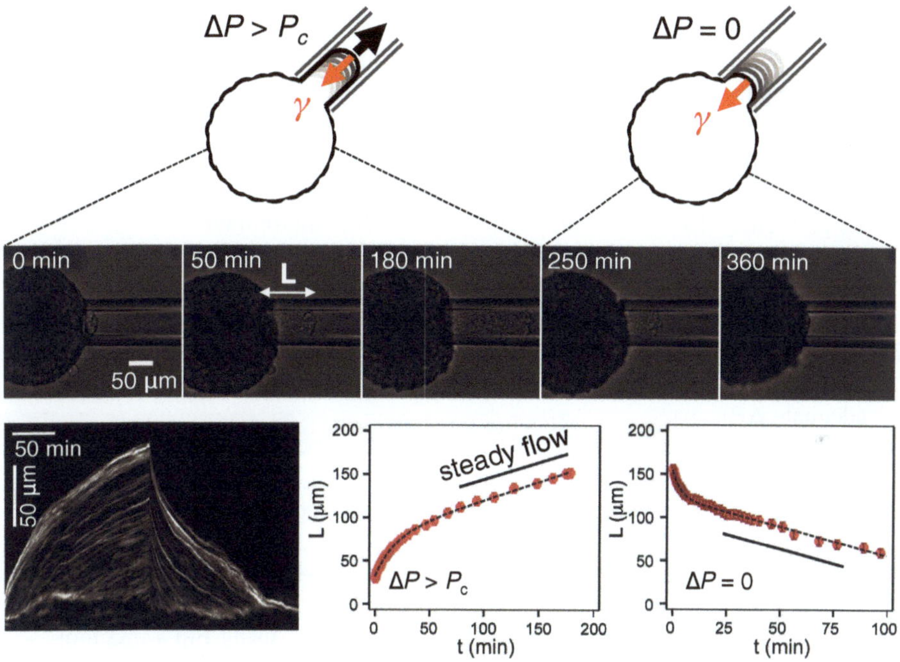

FIGURE 3 Measuring tissue viscosity.

Schematic diagram of the tissue viscosity measurement technique. Using a micropipette of radius R_p, the tissue is aspirated with a pressure ΔP, which is above the critical pressure P_c. After elastic stresses are dissipated, the rate of aspiration $\dot{L}_{asp}$ is limited by the surface tension γ and viscosity η of the tissue. When the pressure in the micropipette is set to zero $\Delta P = 0$, the tissue retracts under the action of its surface tension γ at a rate $\dot{L}_{ret}$, which is limited by the tissue viscosity η. Pictures of aspiration and retraction time-series of an aggregate of cells (Movie 2) and the corresponding kymograph showing the aspiration-retraction dynamics. The aspiration pressure ΔP is 1.5 kPa. The length of the aspirated tissue, L, is measured over time and plotted on the graph below (Guevorkian et al., 2010). The rate of aspiration $\dot{L}_{asp}$ and retraction $\dot{L}_{ret}$ are obtained from the slope of the linear part of the graph.

obtained when the tissue is aspirated to hemisphere or, alternatively, from the aspiration and retraction flow rates: $P_c = \Delta P \frac{\dot{L}_{ret}}{\dot{L}_{asp} + \dot{L}_{ret}}$. The surface tension of spherical aggregate can then be calculated from the Young–Laplace equation as before, $\gamma = P_c/2(1/R_p - 1/R_0)$, where R_0 is the radius of curvature at rest. For tissues of large dimensions, this relationship can be approximated to $\gamma = P_c/2R_p$.

3.2 MATERIALS

- Cell aggregates in suspension
- Cell culture medium
- Mineral oil for prolonged experiments
- Nonadhesive glass bottom culture dish
- Micropipette typically at least three times larger than the cells and one-third of the aggregate size

3.3 METHODS

1. As previously described, mount and calibrate the micropipette.
2. Using the micropipette, choose and orient the tissue before getting a hold of it with a low pressure (20–30 Pa) that is sufficient to enable raising the tissue from the bottom of the dish.
3. Apply an aspiration pressure ΔP larger than the critical pressure P_c above which the tissue starts flowing inside the micropipette (see Section 2.4 for finding the adequate ΔP).
4. Record the entry of the aggregate into the micropipette by taking images every few seconds.
5. Once a sufficient amount of tissue has entered the micropipette (see Section 2.4), set the pressure to zero.
6. Record the retraction of the aggregate by taking images every few seconds.
7. Measure the radius of the micropipette opening R_p and the length of aspirated tissue inside the micropipette over time [for example, using Fiji's line tool (Schindelin et al., 2012)].
8. Calculate the rate of aspiration $\dot{L}_{asp}$ and retraction $\dot{L}_{ret}$, which corresponds to the slopes of the linear parts of the aspiration-retraction curves (Fig. 3).
9. Calculate the tissue viscosity η using $\eta = \frac{R_p \Delta P}{3\pi\,(\dot{L}_{asp} + \dot{L}_{ret})}$.

3.4 DISCUSSION

Other techniques have been developed to evaluate tissue rheology, such as parallel plate compressor (Forgacs, Foty, Shafrir, & Steinberg, 1998; Foty, Pfleger, Forgacs, & Steinberg, 1996) or centrifugation (Ninomiya & Winklbauer, 2008; Phillips & Steinberg, 1978). Similarly to these techniques, the microaspiration method described above is a bulk measurement approach giving an average viscosity of the tissue, which is considered homogeneous (Guevorkian et al., 2010). The notable benefit of the microaspiration technique is that it can be applied to tissues of irregular shape and opens the possibility to measure the mechanical properties of tissues in vivo (Majkut et al., 2013; Porazinski et al., 2015). Moreover, microaspiration can be easily combined with any inverted microscope allowing the high-resolution imaging of cellular rearrangements during the deformation. This will be instrumental to better understand the dynamics of cellular rearrangements during morphogenesis. In addition, similarly to parallel plate compression, tissue surface tension can be measured simultaneously in the case of spherical tissues. This is crucial to understand how tissues change shape over time and how tissues of different surface tension arrange themselves.

A few preliminary aspirations will be required to find the adequate aspiration pressure ΔP and time of release of the pressure. If the aspiration pressure ΔP is lower than P_c, the tissue will not flow inside the micropipette. If the aspiration pressure ΔP is above but close to P_c, the tissue will flow very slowly, thereby limiting the throughput of the experiment. On the other hand, if the aspiration pressure ΔP is too high, the tissue will flow rapidly but may get damaged. Typically, for tissues with low cell—cell adhesion, 5—10 min of aspiration are enough to determine the tissue viscosity and surface tension, whereas a couple of hours may be needed for strongly adhering tissues.

GENERAL CONCLUSIONS

Microaspiration has been used to manipulate and probe living materials for decades. The contribution of this method to our modern understanding of cell and tissue mechanics is invaluable. This knowledge now allows us to bring microaspiration to more intricate contexts observed in vivo. The surface tension mapping and viscosity measurement methods described above will help characterize native morphogenetic processes. These two methods can be used complementarily to obtain information not only at the cell and tissue level but also at the static and dynamic levels. Moreover, applying those methods after alteration of key molecular components will bring key mechanistic insights into these morphogenetic processes.

As a final note, a microaspiration setup can also be used to manipulate and perturb biological processes (Kee & Robinson, 2013) or to probe cell—cell adhesion in vitro (Biro & Maître, 2015; Maître et al., 2012). Therefore, microaspiration setups can be valuable and versatile tools for laboratories interested in the shaping of cells and tissues in the context of development and disease.

SUPPLEMENTARY DATA

Supplementary data related to this article can be found online at http://dx.doi.org/10.1016/bs.mcb.2016.11.012.

REFERENCES

Behrndt, M., Salbreux, G., Campinho, P., Hauschild, R., Oswald, F., Roensch, J., … Heisenberg, C.-P. (2012). Forces driving epithelial spreading in zebrafish gastrulation. *Science (New York, NY), 338*, 257–260.

Biro, M., & Maître, J.-L. (2015). Dual pipette aspiration: a unique tool for studying intercellular adhesion. *Methods in Cell Biology, 125*, 255–267.

Charras, G., & Paluch, E. (2008). Blebs lead the way: how to migrate without lamellipodia. *Nature Reviews Molecular Cell Biology, 9*, 730–736.

Collinet, C., Rauzi, M., Lenne, P.-F., & Lecuit, T. (2015). Local and tissue-scale forces drive oriented junction growth during tissue extension. *Nature Cell Biology, 17*, 1247–1258.

Dai, J., Ting-Beall, H., Hochmuth, R., Sheetz, M., & Titus, M. (1999). Myosin I contributes to the generation of resting cortical tension. *Biophysical Journal, 77*, 1168–1176.

von Dassow, M., & Davidson, L. (2009). Natural variation in embryo mechanics: gastrulation in *Xenopus laevis* is highly robust to variation in tissue stiffness. *Developmental Dynamics: An Official Publication of the American Association of Anatomists, 238*, 2–18.

Evans, E., & Yeung, A. (1989). Apparent viscosity and cortical tension of blood granulocytes determined by micropipet aspiration. *Biophysical Journal, 56*, 151–160.

Forgacs, G., Foty, R., Shafrir, Y., & Steinberg, M. (1998). Viscoelastic properties of living embryonic tissues: a quantitative study. *Biophysical Journal, 74*, 2227–2234.

Foty, R. A., Pfleger, C. M., Forgacs, G., & Steinberg, M. S. (1996). Surface tensions of embryonic tissues predict their mutual envelopment behavior. *Development (Cambridge, England), 122*, 1611–1620.

Gonzalez-Rodriguez, D., Guevorkian, K., Douezan, S., & Brochard-Wyart, F. (2012). Soft matter models of developing tissues and tumors. *Science (New York, NY), 338*, 910–917.

Guck, J., Schinkinger, S., Lincoln, B., Wottawah, F., Ebert, S., Romeyke, M., … Bilby, C. (2005). Optical deformability as an inherent cell marker for testing malignant transformation and metastatic competence. *Biophysical Journal, 88*, 3689–3698.

Guevorkian, K., Colbert, M.-J., Durth, M., Dufour, S., & Brochard-Wyart, F. (2010). Aspiration of biological viscoelastic drops. *Physical Review Letters, 104*, 218101.

Heisenberg, C.-P., & Bellaïche, Y. (2013). Forces in tissue morphogenesis and patterning. *Cell, 153*, 948–962.

Kee, Y.-S., & Robinson, D. N. (2013). Micropipette aspiration for studying cellular mechanosensory responses and mechanics. *Methods in Molecular Biology (Clifton, NJ), 983*, 367–382.

Larson, S. M., Lee, H. J., Hung, P.-H., Matthews, L. M., Robinson, D. N., & Evans, J. P. (2010). Cortical mechanics and meiosis II completion in mammalian oocytes are mediated by myosin-II and Ezrin-Radixin-Moesin (ERM) proteins. *Molecular Biology of the Cell, 21*, 3182–3192.

Levayer, R., Hauert, B., & Moreno, E. (2015). Cell mixing induced by myc is required for competitive tissue invasion and destruction. *Nature, 524*, 476–480.

Maître, J.-L., Berthoumieux, H., Krens, S. F. G., Salbreux, G., Jülicher, F., Paluch, E., … Heisenberg, C.-P. (2012). Adhesion functions in cell sorting by mechanically coupling the cortices of adhering cells. *Science (New York, NY), 338*, 253–256.

Maître, J.-L., Niwayama, R., Turlier, H., Nédélec, F., & Hiiragi, T. (2015). Pulsatile cell-autonomous contractility drives compaction in the mouse embryo. *Nature Cell Biology, 17*, 849–855.

Maître, J.-L., Turlier, H., Illukkumbura, R., Eismann, B., Niwayama, R., Nédélec, F., … Hiiragi, T. (2016). Asymmetric division of contractile domains couples cell positioning and fate specification. *Nature, 536*, 344–348.

Majkut, S., Idema, T., Swift, J., Krieger, C., Liu, A., & Discher, D. E. (2013). Heart-specific stiffening in early embryos parallels matrix and myosin expression to optimize beating. *Current Biology, 23*, 2434–2439.

Mitchison, J. M., & Swann, M. M. (1954). The mechanical properties of the cell surface I. The cell elastimeter. *Journal of Experimental Biology, 31*, 443–460.

Mitchison, J. M., & Swann, M. M. (1955). The mechanical properties of the cell surface III. The sea-urchin egg from fertilization to cleavage. *Journal of Experimental Biology, 49*, 734–750.

Ninomiya, H., & Winklbauer, R. (2008). Epithelial coating controls mesenchymal shape change through tissue-positioning effects and reduction of surface-minimizing tension. *Nature Cell Biology, 10*, 61–69.

Phillips, H. M., & Steinberg, M. S. (1978). Embryonic tissues as elasticoviscous liquids. I. Rapid and slow shape changes in centrifuged cell aggregates. *Journal of Cell Science, 30*, 1–20.

Porazinski, S., Wang, H., Asaoka, Y., Behrndt, M., Miyamoto, T., Morita, H., … Furutani-Seiki, M. (2015). YAP is essential for tissue tension to ensure vertebrate 3D body shape. *Nature*, 1–16.

Rauzi, M., Krzic, U., Saunders, T. E., Krajnc, M., Ziherl, P., Hufnagel, L., … Leptin, M. (2015). Embryo-scale tissue mechanics during *Drosophila* gastrulation movements. *Nature Communications, 6*, 8677.

Schindelin, J., Arganda-Carreras, I., Frise, E., Kaynig, V., Longair, M., Pietzsch, T., … Cardona, A. (2012). Fiji: an open-source platform for biological-image analysis. *Nature Methods, 9*, 676–682.

Sugimura, K., Lenne, P. F., & Graner, F. (2016). Measuring forces and stresses in situ in living tissues. *Development (Cambridge, England), 143*, 186–196.

Swift, J., Ivanovska, I. L., Buxboim, A., Harada, T., Dingal, P. C. D. P., Pinter, J., … Discher, D. E. (2013). Nuclear lamin-A scales with tissue stiffness and enhances matrix-directed differentiation. *Science (New York, NY), 341*, 1240104.

Tinevez, J.-Y., Schulze, U., Salbreux, G., Roensch, J., Joanny, J.-F., & Paluch, E. (2009). Role of cortical tension in bleb growth. *Proceedings of the National Academy of Sciences of the United States of America, 106*, 18581–18586.

Wolfenson, H., Meacci, G., Liu, S., Stachowiak, M. R., Iskratsch, T., Ghassemi, S., … Sheetz, M. P. (2015). Tropomyosin controls sarcomere-like contractions for rigidity sensing and suppressing growth on soft matrices. *Nature Cell Biology, 18*, 33–42.

CHAPTER

In-vivo analysis of morphogenesis in plants

11

T. Stanislas, O. Hamant, J. Traas[1]

Laboratoire de Reproduction et Développement des Plantes, ENS-Lyon, INRA, CNRS, UCBL, Université de Lyon, 46 allée d'Italie, 69364 Lyon Cedex 07, France

[1]*Corresponding author: E-mail: Jan.Traas@ens-lyon.fr*

CHAPTER OUTLINE

Methods in Cell Biology, Volume 139, ISSN 0091-679X, http://dx.doi.org/10.1016/bs.mcb.2016.11.008

Abstract

Although many molecular regulators of morphogenesis have been identified in plants, it remains largely unknown how the molecular networks influence local cell shape and how cell growth, form, and position are coordinated during tissue and organ formations. So far, analyses of gene function in morphogenesis have mainly focused on the qualitative analysis of phenotypes, often providing limited mechanistic insight into how particular factors act. For this reason, there has been a growing interest in mathematical and computational models to formalize and test hypotheses. These require much more rigorous, quantitative approaches; in parallel, new quantitative and correlative imaging pipelines have been developed to study morphogenesis. Here, we describe a number of such methods, focusing on live imaging.

INTRODUCTION

Whereas many molecular regulators of morphogenesis have been identified in both animals and plants, it remains largely unknown how the molecular networks influence local cell shape and how cell growth, form, and position are coordinated during tissue and organ formations. So far, analyses of gene function in morphogenesis have mainly focused on the qualitative analysis of phenotypes, often providing limited mechanistic insight into how particular factors act. For this reason, there has been a growing interest in mathematical and computational models to formalize and test hypotheses. These require much more rigorous, quantitative approaches; in parallel, new quantitative and correlative imaging pipelines have been developed to study morphogenesis. Here, we describe a number of such pipelines, focusing on plants.

If considered at a sufficiently small scale, every local shape change in a developing organ can be decomposed into a set of regional transformations corresponding to (1) the local growth rate and two parameters linked to growth anisotropy, namely (2) the degree at which a region grows in a particular direction and (3) the growth directions themselves (e.g., Coen, Rolland-Lagan, Matthews, Bangham, & Prusinkiewicz, 2004). By regulating these three parameters independently during growth, local gene action thus leads to specific changes in shape. This implies that, to understand the molecular control of development, one needs to specify gene function in terms of *quantitative* local changes in geometry over time. Two important factors have allowed for a transition toward truly quantitative approaches. First, as plant cells are tightly glued together by their external matrix, plant morphogenesis does not involve cell migration. In addition, cells in most young plant tissues do not disappear from the growing tissues through cell death. As a result, cells and their descendants maintain their relative positions in a growing tissue. This makes it much easier to follow local growth rates and directions than for example in animal systems. Second, technical improvements in the field of life imaging have made it possible to follow plant growth in vivo with great spatial and temporal resolutions.

Even though the quantitative analysis of growth is a necessary step, it is not sufficient for a full understanding of the process. This is because the link between genes and shape is obviously not direct, and from a mechanistic point of view, it is crucial to involve the cell mechanical features that are determinantal for the plant mode of development. In plants, osmotic water uptake leads to a very high internal hydrostatic pressure, turgor pressure, which is counteracted by the presence of a stiff extracellular matrix or cell wall. Synthesis and remodeling of this wall allows the cell to yield to turgor pressure, thus causing the cell to grow (for reviews see e.g., Cosgrove, 2014; Wolf et al. 2012; for a theoretical analysis, see Ali & Traas, 2016). The physical properties of the wall largely depend on the presence of stiff cellulose microfibrils, cross-linked via a matrix of polysaccharides and proteins (Cosgrove, 2015, 2014). These microfibrils are often deposited in highly ordered arrays, channeling the direction of cell growth, while their synthesis and degree of cross-linking influences growth rate. Understanding how plant morphogenesis is coordinated therefore also comes down to understanding how molecular regulatory networks interact with cell wall synthesis, wall stiffness, and mechanical anisotropy. This has led to the development of methods to analyze changes in the cell wall composition and texture, in situ and even in vivo.

In the following paragraphs, we will present a number of in-vivo methods that have been developed to study growth and wall properties during morphogenesis. Hereby, we will mainly focus on methods adapted for Arabidopsis, which has been most widely studied. The methods described here are centered on meristematic tissues, where multiple morphogenetic events take place. Note that they have been adapted to other growing tissues and organs, such as hypocotyls, leaves, or stems. Finally, morphogenesis is a complex process, with multiple feedbacks and operating at multiple scales. Notably, gene activity depends on feedback from geometry [either via biochemical (e.g., Cui, Copsey, Green, Bangham, & Coen, 2010; Jaeger, Irons, & Monk, 2008) or mechanical cues (Farge, 2003; Landrein et al., 2015)] and gene input is itself influenced by shape. To address this complexity, modeling approaches have been developed. We will not provide detailed information on these models, but give a general overview in the last part of this chapter as a basis for further reading.

1. MEASURING THE GROWTH: LIVE IMAGING OF GROWING TISSUES AT CELLULAR RESOLUTION

1.1 IMAGING THE GROWTH IN MERISTEMATIC TISSUES

We describe here a methodology using confocal microscopy to observe the cell contours in root apical meristems (RAMs) or shoot apical meristems (SAMs), which is usually done by marking the plasma membrane or cell wall. Two tools are generally used to stain these compartments:

- Plants stably expressing genetically encoded *fluorescent protein* (FP) tags and
- Dyes able to stain either the wall or the plasma membrane (see Table 1).

Table 1 Examples of Fluorescent Reporter Lines or Fluorescent Probes for Cell Contours

	Excitation Lines (nm)	Emission Window (nm)	RAM/SAM	Reference
p35S:EGFP-Lti6b	488	492–650	Y/Y	Cutler, Ehrhard, Griffitts, & Somerville (2000)
pUB10: NPSN12-mCherry (Wave 131)	561	568–700	Y/N	Geldner et al. (2009)
FM4-64	488	620–760	Y/Y	Molecular probes
FM1-43	488	520–620	Y/Y	Molecular probes
Propidium Iodide	561	580–630	Y/Y	Molecular probes

The methods described for the shoot apical meristem are largely based on those described in Grandjean et al. (2004), Reddy, Heisler, Ehrhardt, & Meyerowitz (2004), Heisler et al. (2005) and Fernandez et al. (2010). For roots the first live imaging in Arabidopsis was described more than 20 years ago (van den Berg, Willemsen, Hage, Weisbeek, & Scheres, 1995). Since then, many studies have described variants of this method; here, we will simply describe a general basic approach that does not require specific equipments (e.g., vertical microscopes) or miniaturized growth chambers. Further details, in particular on complementary methods (e.g., staining and immunolabeling of fixed material), can also be found in Žársky and Cvrčková (2014).

1.1.1 Media and stock solutions

1. *Sterilization solution:* This solution is to prepare fresh. Dissolve 88 mg of Dichloro-isocyanuric acid in 1 mL H_2O final and add 9 mL ethanol 95%. Mix. Note that other methods exist.
2. *Liquid Arabidopsis medium:* Dissolve 11.82 g/L of Duchefa "Arabidopsis" medium (DU0742.0025) in water, add 2 mL/L of 1M $Ca(NO_3)_2$, $4H_2O$. Adjust pH to 5.8 with KOH. Autoclave.
   Note: ½ MS (Murashige & Skoog) medium can also be used, with or without 1% sucrose.
3. *Solid Arabidopsis medium:* Prepare liquid Arabidopsis medium and after adjusting the pH, add 1% of agar. Autoclave.
4. *Dissecting medium:* 10 g of agarose in 1 L of water. Autoclave.
5. *Arabidopsis apex culture medium (ACM):* Dissolve 2.2 g/L of Arabidopsis medium (Duchefa Biochimie, MS basal salt mixture), add 1% sucrose, and adjust

pH to 5.8 with KOH. Add 0.8% agarose. Autoclave. Vitamins (1000X vitamin stock solution) and a final concentration of 555 nM of N6-Benzyladenine are added to the medium (at 55–60°C) before pouring in dishes. In our hands, this recipe works better to conserve the stems.

6. *1000X vitamin stock solution*: For 50 mL, dissolve 5 g myoinositol, 0.05 g nicotinic acid, 0.05 g pyridoxine hydrochloride, 0.05 g thiamine hydrochloride, and 0.1 g glycine in Millipore water. Filter the solution at 22 μm under laminar hood, aliquot, and store at −20°C.
7. *N6-benzyladenine (BAP, a synthetic cytokinin):* Dissolve 1 mg/mL of BAP in a small volume of 1N NaOH and adjust to your final volume with water. Filter the BAP stock at 22 μm under laminar hood, aliquot, and store at −20°C.

1.1.2 Growing plants before imaging

To grow plants in vitro, seeds are surface sterilized (1 min in the sterilization solution, remove solution, two times 1 min in ethanol 95%, remove solution, leave under the fume hood overnight to dry), sown onto Arabidopsis or ½ MS medium plates and left for 3 days at 4°C in the dark.

1. *For root meristem visualization*: Seedlings are grown vertically for 5 days on MS plate. The photoperiod is set up to 16-h light/8-h dark at 22°C.
   Note: Other photoperiods can be used but since this will affect root morphology, it is important to compare plants grown under similar photoperiod conditions.
2. *For shoot meristem visualization*:
   Plants can be either grown on soil or in vitro:
   a. Seedlings grown for 1–2 weeks in vitro are transferred on soil (1 plant/pot) and kept at 22°C in short-day conditions (8-h light/16-h dark) for at least 3 weeks, until the rosette covers most of the surface of the pot. Plants are then transferred to long-day conditions (16-h light/8-h dark) and are used when the inflorescence stem is about 2 cm above the rosette leaves.
   b. To facilitate access to the meristem, plants can be grown in vitro in the presence of 10 μM of the auxin transport inhibitor NPA (1-N-Naphthylphthalamic acid) (Grandjean et al. 2003; Reinhardt, Mandele, & Kuhlenmeier, 2000; Sassi et al., 2014) naked meristems are formed, which can be directly visualized. This is preferentially done by growing the individual plants in small sterile plastic boxes (diameter about 4–5 cm, height about 3 cm). Alternatively, 3-cm high petri dishes can be used.

   Note: since it takes at least four weeks before the inflorescence meristem emerges, it is advised to check fluorescence and/or phenotype at the seedling stage, and to put on soil only the right plants.

1.1.3 Preparation and mounting of root meristems

1. Mount the seedlings in water between a 24 × 75-mm slide and a 40 × 24-mm coverslip "sandwich" separated by double-sided tape placed at the edges of the slide (Fig. 1A). Adding double-side tape creates a small chamber and avoids

crushing your sample. Approximately, eight roots can be placed using this setup.

2. Place a drop of distilled water on top of the coverslip and carefully approach the tip of the objective (Fig. 1B).
3. Proceed with image acquisition by confocal laser scanner microscopy (CLSM) immediately after mounting.

   Note:

   a. This setup is well suited for relatively short periods, up to one or two days. For longer periods, special microscopes such as a vertical confocal microscope might be required, and in that case seedlings have to be grown in a sterile environment.

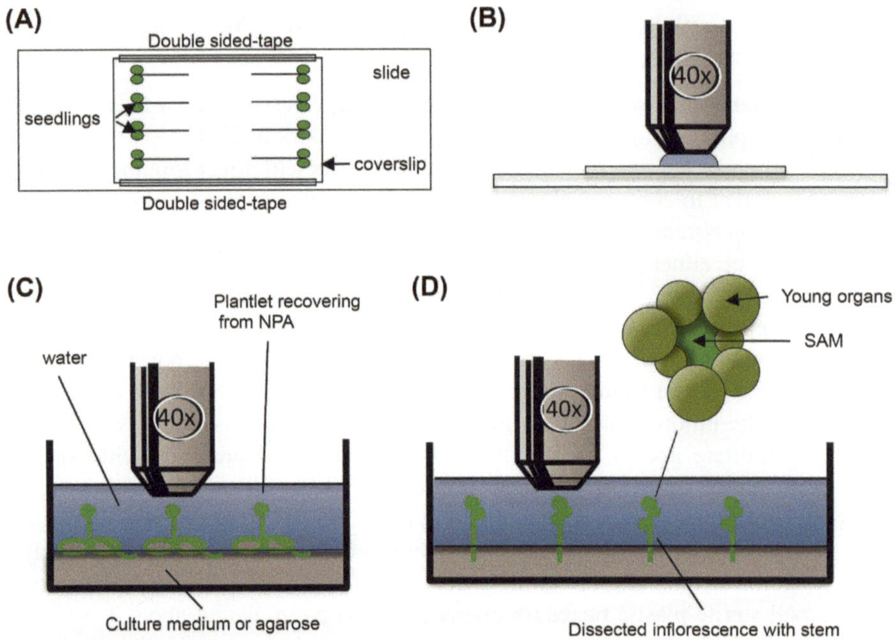

FIGURE 1 Mounting of root and shoot meristems.

(A) Seedlings are mounted in water between slide and coverslip separated by double-sided tape. (B) Seedlings are observed using a water immersion coverslip-corrected objective. (C) Shoot meristems of plants recovering from NPA can directly be observed. Sterile plantlets on NPA with naked, pin-like meristems are transferred to medium without NPA. The meristems will recover after two or three days. (D) Inflorescence meristems are dissected until the shoot apical meristem (in dark green) with young organs is visible. Meristems still attached to short stem segments are placed in a petri dish or box, covered with water and observed with a long-distance water dipping objective.

b. Roots can also be mounted in small rectangular chambers with glass bottom. Young plants are laid down on the glass bottom and subsequently their roots covered with a small block of agar-medium.
c. Plants roots can also be grown directly in vertical boxes or dishes with glass bottom. Roots will grow along the glass and can then directly be observed.

1.1.4 Preparation and mounting of inflorescence meristems

Plants grown in vitro after inhibition of auxin transport with NPA.

1. After 3–4 weeks, about 50% of the plantlets that have been grown on NPA containing medium form naked inflorescence meristems [Note that this proportion depends on both the ecotype and the environment (light, temperature, brand of medium used, ...)].
2. Organs can be induced on these meristems by transferring the plants to normal medium without NPA. Use tall boxes, as the plants will be fully immersed in water during imaging. Add 0.5% Agarose (ideally low melting point) to fix the plants on medium. New organs will then form within 48–96 h.
3. When under the microscope, immerse the plants with water just before observation to avoid dehydration of samples.
4. Use a long-distance water immersion ("dipping") objective (20-40x) on an upright microscope to observe the sample under water. With a 100-μL pipette, flush around the meristem to remove trapped air bubbles at the tip of the lens.
5. Remove the water after observation and put the plants back in the growth chamber for a later observation.

Plants grown on soil.

Inflorescence meristems are covered by older organs, preventing their analysis. It is first necessary to remove these organs.

1. Pour 1% agarose in water (dissecting medium) in round petri dishes (60 × 15 mm) for sample mounting. Prepare multiple dishes. Keep unused dishes at 4°C in a plastic wrap to avoid agarose dessication.
2. Manually remove the larger flowers and siliques from the inflorescence stem. Cut the stem 1 cm under the apex and place it in a dissecting dish under a binocular.
3. Using a pair of sharp forceps (use a sharpening stone to turn your forceps into mini scissors), remove floral buds by pinching where the pedicel meets the stem. Stop when the shoot apical meristem (a glossy dark-green dome) with young organs is visible (Fig. 1C).

Note:

- Dissection can be done under water to avoid dehydration of the sample but it is more challenging since water induces optical aberrations. When dissecting in air, try to be as fast as possible.
- Watering your plants for 1–2 h before dissecting will increase turgor pressure in the meristem and this will help when removing organs.

4. Transfer the dissected inflorescence meristem to a fresh Arabidopsis apex culture medium (ACM), close to the center of the petri dish or box (to ensure that the lens can access the meristem without touching the side of the dish). Push the stem into the ACM medium until the apex is located just above the agar surface.
5. Cover with water to avoid dehydration of samples.
6. Use a long-distance water immersion ("dipping") objective (20-40x) on an upright microscope to observe the sample under water. With a 100-μL pipette, flush around the meristem to remove trapped air bubbles at the tip of the lens.
7. Remove the water after observation and put the plants back in the growth chamber for a later observation.

Note: Dissecting a meristem induces stress. It is thus best to wait several hours or even overnight between dissection and observation.

1.1.5 Staining before imaging

To analyze morphogenesis, it is necessary to visualize tissue anatomy. Direct confocal imaging can be done if the meristem is coming from a plant expressing genetically encoded fluorescent protein fused to a protein/domain/tag localized at the PM (such as *p35S::EGFP-LTI6b* line). Otherwise, fluorescent dyes for either plasma membrane (FM4-64 or FM1-43 (FM)) or cell wall (Propidium Iodide (PI)) can be used (see Table 1). Both staining methods work for the shoot and root meristems. FM4-64 has an emission spectrum in the red (620−750 nm) whereas FM1-43 emits in the green (520-620 nm). Depending on the fluorescent protein of interest in the sample, choose the FM dye accordingly. Note that both PI and FM stains can affect cell function if used at high concentrations.

Staining the plasma membranes or cell walls in roots.

The same procedure is used for plasma membrane and cell wall staining with FM and PI dyes, respectively.

1. Prepare Stock solution FM 1 mg/mL in water, PI 1 mg/mL in water
2. Pipette out 2 mL of liquid Arabidopsis medium into a 12-well plate.
3. Add 2 μL of FM or 10 μL of PI (Working solution: FM 1 μg/mL; PI 5 μg/mL).
4. Collect seedlings, submerge them in the liquid MS medium and incubate for maximum 5 min (on ice for FM staining).
5. Wash twice with liquid MS medium (on ice for FM staining).
6. Mount the sample in liquid MS medium at room temperature for live imaging as described above.
7. Immediately proceed to image acquisition by CLSM.

Staining plasma membranes or cell walls in the SAM.

The same procedure is used for staining with FM and PI dyes:

1. Dilute stock solution (FM 1 mg/mL in water, PI 0.1 mg/mL in water) to a working solution of FM 330 μg/mL and PI 1 mg/mL.

2. To facilitate dye penetration, immerse the meristem for 5 min in water before staining.
3. Put a 2-μL drop of FM working solution or pour the PI solution to the petri dish.
4. Incubate at least 5 min and wash twice with water.
5. Immediately proceed to image acquisition using CLSM.

1.1.6 Imaging

Long-distance water immersion objectives such as HCX IRAPO L (25x) or a water immersion coverslip-corrected Objective PL APO (40x) can be used.

1. Dip the tip of the clean objective in the distilled water above the sample.
2. Air bubbles between your sample and objective can be removed either by raising the objective in and out of the water repeatedly, flushing the bubble with a 100-μL pipette, or by cleaning the objective with lens paper.
3. Acquire a 3D image (a stack) of a biological sample according to the fluorescence characteristics of the fluorophore or dye (Table 1).
   Note: The following can improve the final results:
   - Collect images with 16 bits per channel.
   - Use a Z-step of 0.5 μm.
   - A signal can be quantified only if the image is not saturated in intensity.
   - For a signal used for 3D reconstruction, it can help to oversaturate the cell contour (FM or PI) signal, while leaving the intracellular signal from the fluorophore of interest not saturated.
   - Start the stack a few microns above the sample. This will improve the 3D reconstruction.
   - Generate separated TIF files with one channel per file.

1.2 SEGMENTATION, THREE-DIMENSIONAL RECONSTRUCTIONS, AND QUANTIFICATION OF GROWTH

Different methods and tools exist to segment the images. We will present here only an approach using the MorphoGraphX (MGX) software (Barbier de Reuille et al., 2015; www.MorphoGraphX.org), which is accessible and relatively user-friendly. Other methods such as the MARS (Multi Angle Reconstruction and Segmentation) should also become available for a wider audience (Fernandez et al., 2010).

MGX-based segmentation is performed in several steps. First, the exact surface of the meristem is detected by the level set method, through thresholding of the image. This surface is then smoothened to obtain a high-resolution representation of the outer meristem wall in 3D. MGX is subsequently used to analyze the extracted curved 2D surfaces. The workflow presented here is optimized for meristems stained with FM4-64. For more details, refer to the user manual of MGX (http://www.mpipz.mpg.de/4085950/MGXUserManual.pdf).

1.2.1 Surface detection

Start with a TIFF file (see above), visualized in the MGX software (Fig 2A). After a Gaussian blur of a sigma which is approximately twice the least value in the voxel size (Fig. 2B), the level set is initialized by a threshold (of 10, for example) (Fig. 2C).

Note: If holes appear at this step, decrease the threshold until there are no more holes. The level-set function depends on the parameters. In our hands, Alpha = 0, Beta = 0.1, and Lambda = 10 (Fig. 2D).

Note: Check if the surface found corresponds to your signal. Reopen your original TIFF image Stack/Stack1/Main/open. In the main panel, tick Main and

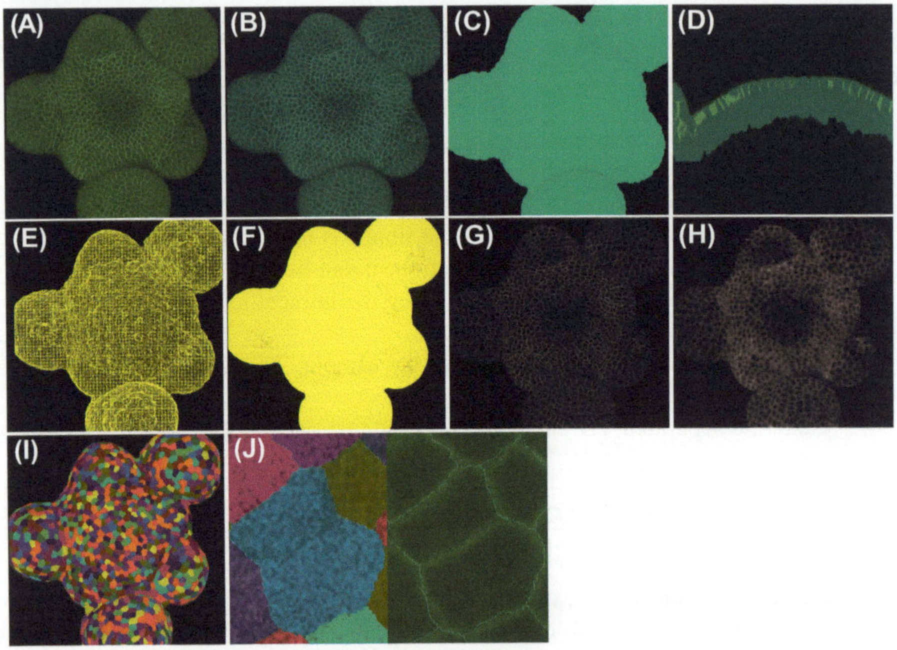

FIGURE 2 Screenshot of the different steps from a 3D image to a cell-segmented 2D curved surface.

(A) A TIFF file is opened using the MorphoGraphX software. (B) The signal is blurred to increase the noise and improve the surface detection. (C) The surface is detected using the level set function. (D) It is important to check that the surface detected is similar to the surface provided by the signal on cross-sections. (E) Create a mesh of the surface and (F) smoothen and subdivide it. (G) A signal is projected on the mesh giving the contour of the cell at the surface of the meristem. (H) To segment the cell, the outlines of the cell are blurred and (I) after auto-seeding and segmentation, cells segmented are visualized. (J) It is important to check the segmentation by comparing with the original plasma membrane or cell wall signal.

Work and in the View panel tick Enabled (Fig. 2E). If the fluorescent signal is not at the surface, change parameters in Steps 4 and 5 below.

1.2.2 Mesh creation and signal projection

1. Create a mesh. Process/Mesh/Creation/Marching Cubes Surface with cube size = 3 μm (half the size of a cell).

   Note: In the main panel, select only the mesh, with a view of all (Fig. 2F)
2. Remove the bottom part of the mesh to reduce the number of vertices.
3. Smoothen the mesh and subdivide the mesh in Process/Mesh/Structure/Smooth Mesh and Subdivide Mesh (Fig. 2G).
4. Project the signal onto the mesh Process/Mesh/Signal/Project Signal with a minimal distance of 0 μm and a maximal distance = 3 μm (Fig. 2H).
   Note: You can either project the signal used to generate the mesh (FM or PI dyes, LTI6b-GFP) or another signal used to analyze protein expression. Since physical properties of plant cell walls are largely dependent on a microtubule organization, it is interesting to project in the curved 2D surfaces the signal of fluorescent proteins associated with microtubules (see below for a discussion on microtubule quantification).
5. Save the mesh, the work, and a screenshot of the projection.

1.2.3 Seeding and segmentation

1. Blur the cell outline with Process/Mesh/Signal/Gaussian Blur with a radius of 2 μm (Fig. 2I).
2. Seed the cell automatically with Process/Mesh/Segmentation/Auto-Seeding with a given radius of 2 μm.
3. Reproject the signal onto the Mesh Process/Mesh/signal/project Signal with the same minimum and maximum distances used in Section 1.2.2 in list 4
4. Blur the cell outlines Process/Mesh/Signal/Gaussian Blur with a radius of 1 μm
5. Use the watershed function to propagate the labels (Fig. 2J) Process/Mesh/Segmentation/Watershed Segmentation
   Note: Check for segmentation errors. Turn the mesh visualization "on" with a Cell view. In the case of an error (Fig. 2K), first remove the label with the "labels-color" tools to clear the current label, then clear the label in the segmented cell with the error. This is done using the "Fill-label" tool. Add new seeds and run again the watershed.

1.2.4 Quantification of growth

Once a surface is segmented into cells, MGX can be used to quantify cell parameters such as cell area, genetic expression and protein localization, and curvature at the cellular or subcellular level (Barbier de Reuille et al., 2015). After the cell segmentation of multiple time points using the plasma membrane or cell wall signal to create the mesh, MGX can provide quantifications for growth rates, directions and anisotropy as well as tissue curvature, division rate, and plane orientation

(Barbier de Reuille et al., 2015; Hervieux et al., 2016; Landrein et al., 2015; Louveaux, Julien, Mirabet, Boudaoud, & Hamant, 2016). Furthermore, one more signal such as green fluorescent protein (GFP) fused to a protein of interest (protein expression pattern) or a fluorescent tag fused to a specific promoter (genetic expression pattern) can then be projected onto the segmented surface (e.g., Landrein et al., 2015). A heatmap can be generated to reveal quantified genetic expression and protein levels at the cellular, or subcellular scale.

2. PROBING THE WALL STIFFNESS AND ANISOTROPY

As indicated in the introduction, quantitative growth analysis is only a first step toward a better understanding of morphogenesis. We also need to obtain information on the structural elements of the cell, in particular the physical properties of the wall. In this context, two in-vivo methods are relevant:

- The quantification of microtubule orientation. Since cortical microtubules usually guide the deposition of cellulose microfibrils, their orientation can serve as a proxy for wall mechanical anisotropy.
- The definition of wall stiffness using atomic force microscopy (AFM) or osmotic treatments.

Since it would be beyond the scope of this chapter to describe AFM in full detail, we will only provide a short description of the setup used (for more details see Milani et al., 2011, 2014; Peaucelle, 2008; Peaucelle et al., 2011 and references therein. For a detailed AFM protocol, see Peaucelle, 2015).

2.1 WALL MECHANICAL ANISOTROPY: LIVE MICROTUBULE IMAGING

2.1.1 Plant lines and observation

Microtubules and microtubule-associated proteins (MAPs) play fundamental roles in plant growth and morphogenesis (see above). Chemical probes were developed to stain microtubules and other cytoskeleton proteins but samples need to be fixed and cell wall permeabilized to allow the penetration of the molecule. To have a dynamic view of microtubules in vivo, it is thus necessary to use plants expressing a genetically encoded fluorescent protein that binds or associates with microtubules. The generation and selection of transgenic A. thaliana expressing fluorescent molecules to visualize microtubules were already described (Lucas, 2013), and many plant lines are already available for observing MTs in living plant cells such as plants overexpressing fluorescent tubulin monomers, MAPs, microtubules-binding domains (MBDs), and EB1, a microtubule plus-end marker (Tirnauera & Biererb, 2000).

It is important to keep in mind that the fluorescence-fusion, and overexpression, of these proteins and domains can interfere with microtubule dynamics. Plants over-expressing the GFP fused with an MBD can exhibit a strong vegetative phenotype (curly and twisted leaves). The transgene can also induce microtubule bundling (Celler et al., 2016). Conversely, so far, no detectable developmental or morphological phenotypes have been observed plants overexpressing GFP-TUA6, where the GFP is fused to the N-terminus of alpha tubulin. It should be noted, however, that this construct can show microtubule instability (for a comparison, see e.g., Louveaux et al., 2016).

It is also important to take into account the promoter used for microtubule observation, especially in the SAM. Indeed, gene expression in this tissue is strictly regulated. When dealing with the SAM, existing lines must be checked for fluorescent signal in the meristem. For more information on the different fluorescent line available, see Celler et al. (2016) for strategies and methods to visualize microtubules in plant cells.

2.1.2 Quantifying microtubule orientations

Different strategies can be used to quantify microtubule orientations. We quantify fibrillar structures in unprocessed microscopy images using FibrilTool. For a detailed protocol describing the tool and how to use it, see (Boudaoud et al., 2014). Briefly, FibrilTool is an ImageJ (http://rsbweb.nih.gov/ij/) macro based on the concept of a nematic tensor. This method does not need any special equipment and is freely accessible. It provides quantifications of microtubules in the form of a bar, the orientation of which represents the average orientation of microtubule array in the ROI, and the length of which is proportional to the array anisotropy. A log file provides the corresponding numbers. One drawback of the method is the need to select an ROI with the Polygon tool (or any other shape tool) to avoid the saturated pixels present at the periphery of the cell (because the cortical microtubule signal accumulates near the cell anticlinal walls, when viewed from the top). This manual region of interest (ROI) selection can be bypassed when using segmented images since it is possible with the MGX-implemented FibrilTool to automatically exclude the saturated pixels at the cell periphery (by defining a distance from the border of the cell that should be excluded). Recently, the nematic tensor concept has been applied to generate a new microtubule-measuring tool, this time without any ROI selection as microtubule array orientation and anisotropy are calculated in circles of constant radius or constant error. This provides subcellular quantifications of microtubule behavior, while also providing an estimation of the error (Tsugawa et al., 2016).

2.2 WALL MECHANICS: CORRELATING EPIFLUORESCENCE SIGNAL WITH NANOINDENTATION

Changes in growth rate involve local changes in the physical properties of cell walls. In particular, growth modulation has been associated with changes in wall elasticity. Note that while growth, being irreversible, would be better quantified with wall plastic properties, there is an accumulating evidence that wall stiffness (or wall elasticity) correlates well with growth rate. To measure this property,

AFM can be used. AFM is a nanoindentation system, using a small cantilever amenable to indent the sample surface very locally. The tip of the cantilever can display various shapes, from 1 to 5-μm wide spherical beads (e.g., Peaucelle et al., 2011; Milani et al., 2013) to measure cell and tissue properties to small pyramids, with a 4-nm radius at their summit (e.g., Milani et al., 2011) to measure the stiffness of the outer wall. Cellular Force Microscopy (CFM) was actually developed, in parallel to commercial AFM techniques, to obtain such larger-scale measurements (RoutierKierzkowska et al., 2012). Beyond the size of the tip, more or less local information on wall stiffness can be obtained depending on indentation depth (Review shrinking the hammer Braybrook). Typically, it is predicted that indentation can deform the outer wall at a distance twice that of indentation, meaning that a 100-nm indentation would affect the wall up to 200-nm deep. When larger indentations are used, the AFM output also reflects turgor pressure in the sample. This can be used to deduce turgor pressure in the sample (Beauzamy et al., 2015; Forouzesh, Goel, Mackenzie, & Turner, 2013; Lintilhac, Wei, Tanguay, & Outwater, 2000; RoutierKierzkowska et al., 2012) or alternatively, AFM measurements can be performed on fully plasmolyzed samples, although this can no longer be considered as true live imaging (e.g., Peaucelle et al., 2011). The system can also be used to obtain detailed images of the surface. Recently, this system has been coupled to a fluorescence microscope to correlate physical properties with gene expression or growth rates. This method, called quantitative tandem epifluorescence and nanoindentation (qTEN), requires the following setup:

1. An AFM mounted under an optical epifluorescence stereomicroscope, with sufficient high resolution to distinguish fluorescent domains. Note that AFMs with a fully open path between the sample and the stereomicroscope objective (e.g., Biocatalyst from Bruker or Nanowizard III from JPK) are best adapted to avoid fluorescent signal loss.
2. A confocal microscope next to the combined AFM-macroscope to visualize the same sample at cellular resolution.

One proceeds as follows:

1. The sample is mounted in a 30-mm petridish and observed under water at room temperature, stabilized in agarose, as described above.
2. To observe shoot meristems, 5-mm stem with a dissected SAM or plantlets regenerating from NPA treatments are used. If necessary, the sample can be further stabilized using 2% agarose in water. Roots are more difficult to measure as the tissue is relatively flexible and can move, in particular, when observed in water.
3. The sample is first observed in the confocal microscope to identify and record specific domains of interest.
4. The 30-mm petri dishes containing the sample is then placed on the XY stage in a specially designed sample holder under the epifluorescence stereomicroscope and the same domain is localized.

5. The AFM head is mounted onto the stage and positioned so that the cantilever is in the vicinity of the GFP-expressing region.
6. Topography and stiffness images are then collected over rectangular areas of typically 30–50 μm per side.
7. Cell shapes in the epidermal cell layer are extracted from the confocal optical stacks, e.g., using MGX.
8. The confocal images and the stiffness maps can then be associated manually based on the cell shapes.
9. Several cycles can be performed, notably to correlate growth or changes in gene expression with wall mechanics (see Milani et al., 2014).
   Note: As mentioned, different tips can be used to probe more or less local mechanical properties. Typically, we use a sharp pyramidal tip and a maximal force of approximately 10 nN (Milani et al., 2011; Sampathkumar et al., 2014), to very locally measure the apparent elastic modulus of the cell wall at user-defined locations.
   Note: For the interpretation of the stiffness measurements, see e.g., Milani et al. (2011, 2014) and references therein.

Because nanoindentation only provides stiffness quantifications across wall thickness, this method is only partially related to growth. Indeed, turgor pressure puts the wall under tension in the wall plane direction, not in the direction perpendicular to the wall. To measure wall properties in the tangential direction, one can apply osmotic treatments, and measure the deformation of the cell contours (as in growth measurements with MGX, see above). This method was notably used to reveal differences in elasticity between the peripheral and central zones of the SAM (Kierzkowski, Lenhard, Smith, & Kuhlemeier, 2013). Interestingly, AFM measurements also provide differences in wall properties between the central zone and the peripheral zone (Milani et al. 2014), suggesting that indentation normal to the surface still provides relevant information on mechanical wall properties and cell growth.

CONCLUSIONS AND PERSPECTIVES: BEYOND QUANTITATIVE LIVE IMAGING

The methods described above allow the user to quantify a range of variables, including cell size, shape, growth rate, growth direction, tissue curvature, gene expression, cytoskeleton behavior, or wall stiffness. How should these be interpreted and how to obtain an integrated view? In the following paragraphs, we will briefly review a number of studies that have addressed these questions. Hereby, we will first highlight work aimed at measuring and correlating several variables on the same sample. Next, we will discuss how modeling is used to obtain an integrated view of these systems.

CORRELATIVE MICROSCOPY

The first example concerns the correlation between cortical microtubule organization and local curvature of the cells and tissues. It was previously proposed that microtubules align along the predicted maximal tension, as prescribed by tissue curvature, in the boundary domain of the SAM. Live imaging of meristems expressing a microtubule marker, and microtubule orientations (with FibrilTool quantifications) were used to test whether this was also the case in other parts of the meristem. Microtubule orientations were correlated with Gaussian curvature measurements of living meristems (measured with the replica method, see e.g., Kwiatkowska & Dumais, 2003). It appeared that while the correlation was high in the boundary, it became much weaker in the other parts of the meristem, suggesting that other factors may contribute to microtubule behavior in the SAM. Conversely, mechanical conflicts associated with growth heterogeneities suggested that the supracellular alignment of microtubules in the boundary domain of the meristem requires both high differential growth rates and negative Gaussian curvature (Burian et al., 2013; Uyttewaal et al., 2012).

Live imaging of the shoot apical meristem also provided quantified data on tissue curvature and gene expression (*STM* promoter activity). Interestingly, a strong correlation between negative Gaussian curvature and *STM* expression was retrieved, notably in the boundary domain of the SAM. This suggested a role of tensile stress (which is indirectly related to curvature) in *STM* expression, a hypothesis further validated by applying different mechanical perturbations (Landrein et al., 2015).

As we have seen, the coupling between fluorescence and AFM measurement now provides the unique opportunity to relate the presence of effectors or gene expression to the mechanical status of individual cells and cell walls. Such a protocol was recently used to correlate wall stiffness in the SAM to *CLV3* expression. In particular, the *CLV3*-expressing domain, containing slowly growing cells, was found to be stiffer than the fast-growing peripheral zone. In young flowers, this could also be analyzed dynamically: young floral buds do not express *CLV3* and exhibit a homogeneous stiffness pattern, and as floral buds become older, they finally express *CLV3* and exhibit a stiffer central zone (Milani et al., 2014).

MECHANICAL MODELS OF GROWTH

The data complexity, often covering multiple scales, makes it difficult to interpret the results and to propose pertinent hypotheses. For this reason, models integrating biochemical, physical, and geometrical parameters are becoming more and more important. Different types of models have been developed at different resolutions. The generation of these models, which often take the form of virtual tissues, has been facilitated by the fact that plant cells are glued to their neighbors by the cell wall. As a result, their relative positions are maintained during growth and models (virtual tissues) do not have to take into account cell migration.

A number of models are based on our current views of how growth is controlled in living plants. As mentioned above, the regulation and coordination of growth rates and directions is achieved through the action of the molecular regulatory networks that control the composition and mechanical properties of the cell wall, which yields to the internal turgor pressure in each cell.

To reduce complexity, models have been initially restricted to two dimensions either in the plane or on two-dimensional surfaces folded in three dimensions. To account for the mechanical properties, cell walls can be represented as linear springs, with a particular stiffness and restlength. Growth is obtained by modifying the restlength of the springs, and growth rates can be modulated for instance by changing their stiffness (e.g., Coen et al. 2004; Hamant et al., 2008). By quantifying the leaf growth at the level of the entire organ in Antirhinum and by including markers of cell division as a proxy for growth rate, Nath et al. were able to show that in the *cin* mutant, leaf edges are growing too fast, when compared to the WT. Using a mass spring model, this differential growth was shown to be in principle sufficient to generate leaf ruffling as in the absence of cell movement, and the only way to relax mechanical conflicts is to buckle (Coen et al., 2004; Nath, Crawford, Carpenter, & Coen, 2003; see also Dumais, 2007). This also provided a key conclusion: leaves are not flat by default, but rather ruffled when growth is not tightly controlled.

As we have seen above, methods to image and segment the complete volume of multicellular tissues in 3D at cellular resolution, combined with the measurement of cell mechanics, now provide information regarding the interactions between cells in three dimensions. This led to the generation of models, able to account for the genetic regulation of biophysical processes in 3D multicellular systems. These models use finite element method (FEM) approaches, that is, surfacic elements that allow the modeler to express heterogeneous and/or anisotropic mechanical properties in each wall (Bassel & Smith, 2016). Embryo development was modeled with FEM to underline the interplay between cell geometry and genetic regulators of growth (Bassel et al., 2014). The impact of wall tension on PIN1 polarity was also explored experimentally and in such models (Heisler et al., 2010; Nakayama et al., 2012). FEM models were also used to predict tensile stress patterns in pavement cell walls and correlate them to cortical microtubule behavior (Sampathkumar et al., 2014). Last, different hypotheses for flower morphogenesis were tested at the SAM using FEM (Boudon et al., 2015; Sassi et al., 2014).

In addition to the models with cellular resolution, Coen and colleagues have proposed a framework for tissue growth termed the Growing Polarized Tissue (GPT) framework. This framework does not take into account single cells, but captures the overall growth rates and directions of tissues in three dimensions while taking into account mechanical interactions between different regions. Gene function is expressed in relatively abstract, geometrical terms, specifying the local growth rates and directions without detailing how this is achieved. With this method, the authors were able to propose hypotheses for the genetic regulation of organ formation in different species (e.g., Cui et al. 2010).

A major prospect for the integration of live imaging data in mechanical models is the inclusion of the larger gene network in such models. As of now, gene networks are modeled as massive sets of data but often lack spatialization, or relation to the mechanics of growth. Conversely, mechanical models of growth are integrative in essence, but often rely on a limited set of observed parameters (e.g., microtubule orientations, stiffness of the outer wall, and gene by gene expression patterns). Merging the omics and live imaging is under way, notably thanks to single-cell transcriptomics (which can be done at the last time point of kinetics). The coupling between live imaging and other imaging tools (e.g., Raman spectroscopy for cell wall composition) may also help account for the complexity and diversity of cell effectors in growth. Progress in our understanding of cell wall architecture and chemistry will be instrumental in (1) producing realistic mechanical models, notably including the dynamics of each wall component in relation to wall tension (see Ali & Traas, 2016; Proseus and Boyer, 2008), (2) coding the gene network output into the activity of growth effectors (see e.g., Lin et al., 2013), (3) in further exploring the impact of growth effectors and growth on the gene network (e.g., Abu-Abied et al., 2015).

ACKNOWLEDGMENTS

TS and OH were funded by the ERC grant 'Mechanodevo', JT was funded by the ERC grant 'Morphodynamics'.

REFERENCES

Abu-Abied, M., Mordehaev, I., Sunil-Kumar, G. B., Ophir, R., Wasteneys, G. O., & Sadot, E. (2015). Analysis of microtubule-associated-proteins during IBA-mediated adventitious root induction reveals KATANIN dependent and independent alterations of expression patterns. *PLos One, 10*, e0143828.

Ali, O., & Traas, J. (2016). Force-driven polymerization and turgor-induced wall expansion. *Trends in Plant Science, 21*, 398–409.

Barbier de Reuille, P., Routier-Kierzkowska, A. L., Kierzkowski, D., Bassel, G. W., Schüpbach, T., Tauriello, G., … Smith, R. S. (2015). MorphoGraphX: a platform for quantifying morphogenesis in 4D. *Elife, 4*, e05864.

Bassel, G. W., & Smith, R. S. (2016). Quantifying morphogenesis in plants in 4D. *Current Opinion in Plant Biology, 29*, 87–94.

Bassel, G. W., Stamm, P., Mosca, G., Barbier de Reuille, P., Gibbs, D. J., Winter, R., … Smith, R. S. (2014). Mechanical constraints imposed by 3D cellular geometry and arrangement modulate growth patterns in the Arabidopsis embryo. *Proceedings of the National Academy of Sciences, 111*, 8685–8690.

Beauzamy, L., Louveaux, M., Hamant, O., & Boudaoud, A. (2015). Mechanically, the shoot apical meristem of Arabidopsis behaves like a shell inflated by a pressure of about 1 MPa. *Frontiers in Plant Science, 26*, 1038.

van den Berg, C., Willemsen, V., Hage, W., Weisbeek, P., & Scheres, B. (1995). Cell fate in the Arabidopsis root meristem determined by directional signalling. *Nature, 378*(6552), 62–65.

Boudaoud, A., Burian, A., Borowska-Wykret, D., Uyttewaal, M., Wrzalik, R., Kwiatkowska, D., … Hamant, O. (2014). FibrilTool, an ImageJ plug-in to quantify fibrillar structures in raw microscopy images. *Nature Protocoles, 9*, 457–463.

Boudon, F., Chopard, J., Ali, O., Gilles, B., Hamant, O., Boudaoud, A., … Godin, C. (2015). A computational framework for 3D mechanical modeling of plant morphogenesis with cellular resolution. *PLos Computational Biology, 11*, e1003950.

Burian, A., Ludynia, M., Uyttewaal, M., Traas, J., Boudaoud, A., Hamant, O., … Kwiatkowska, D. (2013). A correlative microscopy approach relates microtubule behaviour, local organ geometry, and cell growth at the Arabidopsis shoot apical meristem. *Journal of Experimental Botany, 64*, 5753–5767.

Celler, K., Fujika, M., Kawamura, E., Ambrose, C., Herburger, K., Holzinger, A., & Wasteneys, G. O. (2016). Microtubules in plant cells: strategies and methods for immunofluorescence, transmission electron microscopy, and live cell imaging. *Methods in Molecular Biology, 1365*, 155–814.

Coen, E., Rolland-Lagan, A. G., Matthews, M., Bangham, J. A., & Prusinkiewicz, P. (2004). The genetics of geometry. *Proceedings of the National Academy of Sciences, 101*, 4728–4735.

Cosgrove, D. J. (2014). Re-constructing our models of cellulose and primary cell wall assembly. *Current Opinion in Plant Biology, 22*, 122–131.

Cosgrove, D. J. (2015). Plant expansins: diversity and interactions with plant cell walls. *Current Opinion Plant Biology, 25*, 162–172.

Cui, M. L., Copsey, L., Green, A. A., Bangham, J. A., & Coen, E. (2010). Quantitative control of organ shape by combinatorial gene activity. *PLos Biology, 8*, e1000538.

Cutler, S. R., Ehrhard, D. W., Griffitts, J. S., & Somerville, C. R. (2000). Random GFP:cDNA fusions enable visualization of subcellular structures in cells of Arabidopsis at high frequency. *Proceedings of the National Academy of Sciences, 97*, 3718–3723.

Dumais, J. (2007). Can mechanics control pattern formation in plants? *Current Opinion in Plant Biology, 10*, 58–62.

Farge, E. (2003). Mechanical induction of Twist in the Drosophila foregut/stomodeal primordium. *Current Biology: CB, 13*(16), 1365–1377 (2008).

Fernandez, R., Das, P., Mirabet, V., Traas, J., Verdeil, J. L., Malandain, G., & Godin, C. (2010). Imaging plant growth in 4D: robust tissue reconstruction and lineaging at cell resolution. *Nature Methods, 7*, 547–553.

Forouzesh, E., Goel, A., Mackenzie, S. A., & Turner, J. A. (2013). In vivo extraction of Arabidopsis cell turgor pressure using nanoindentation in conjunction with finite element modeling. *Plant Journal, 73*, 509–520.

Geldner, N., Dénervaud-Tendon, V., Hyman, D. L., Mayer, U., Stierhof, Y. D., & Chory, J. (2009). Rapid, combinatorial analysis of membrane compartments in intact plants with a multicolor marker set. *Plant Journal, 59*, 169–178.

Grandjean, O., Vernoux, T., Laufs, P., Belcram, K., Mizukami, Y., & Traas, J. (2004). In vivo analysis of cell division, cell growth, and differentiation at the shoot apical meristem in Arabidopsis. *The Plant Cell, 16*, 74–87.

Hamant, O., Heisler, M. G., Jönsson, H., Krupinski, P., Uyttewaal, M., Bokov, P., … Traas, J. (2008). Developmental patterning by mechanical signals in Arabidopsis. *Science, 322*, 1650–1655.

Heisler, M. G., Hamant, O., Krupinski, P., Uyttewaal, M., Ohno, C., Jönsson, H., … Meyerowitz, E. M. (2010). Alignment between PIN1 polarity and microtubule orientation in the shoot apical meristem reveals a tight coupling between morphogenesis and auxin transport. *Plos Biology, 8*, e1000516.

Heisler, M. G., Ohno, C., Das, P., Sieber, P., Reddy, G. V., Long, J. A., & Meyerowitz, E. M. (2005). Patterns of auxin transport and gene expression during primordium development revealed by live imaging of the Arabidopsis inflorescence meristem. *Current Biology, 15*, 1899–1911.

Hervieux, N., Dumond, M., Sapala, A., Routier-Kierzkowska, A. L., Kierzkowski, D., Roeder, A. H., … Hamant, O. (2016). A mechanical feedback restricts sepal growth and shape in Arabidopsis. *Current Biology, 26*, 1019–1028.

Jaeger, J., Irons, D., & Monk, N. (2008). Regulative feedback in pattern formation: towards a general relativistic theory of positional information. *Development, 135*(19), 3175–3183.

Kierzkowski, D., Lenhard, M., Smith, R., & Kuhlemeier, C. (2013). Interaction between meristem tissue layers controls phyllotaxis. *Developmental Cell, 26*, 616–628.

Kwiatkowska, D., & Dumais, J. (2003). Growth and morphogenesis at the vegetative shoot apex of Anagallis arvensis L. *Journal of Experimental Botany, 54*, 1585–1595.

Landrein, B., Kiss, A., Sassi, M., Chauvet, A., Das, P., Cortizo, M., … Hamant, O. (2015). Mechanical stress contributes to the expression of the STM homeobox gene in Arabidopsis shoot meristems. *Elife, 4*, e07811.

Lin, D., Cao, L., Zhou, Z., Zhu, L., Ehrhardt, D., Yang, Z., & Fu, Y. (2013). Rho GTPase signaling activates microtubule severing to promote microtubule ordering in Arabidopsis. *Current Biology, 23*, 290–297.

Lintilhac, P. M., Wei, C., Tanguay, J. J., & Outwater, J. O. (2000). Ball tonometry: a rapid, nondestructive method for measuring cell turgor pressure in thin-walled plant cells. *Journal of Plant Growth Regulation, 19*, 90–97.

Louveaux, M., Julien, J. D., Mirabet, V., Boudaoud, A., & Hamant, O. (2016). Cell division plane orientation based on tensile stress in Arabidopsis thaliana. *Proceedings of the National Academy of Sciences,, 113*, E4294–E4303.

Lucas, J. (2013). Live-cell imaging of microtubules and microtubule-associated proteins in *Arabidopsis thaliana. Methods in Cell Biology, 115*, 231–246.

Milani, P., Braybrook, S. A., & Boudaoud, A. (2013). Shrinking the hammer: micromechanical approaches to morphogenesis. *Journal of Experimental Botany, 64*, 4651–4662.

Milani, P., Gholamirad, M., Traas, J., Arnéodo, A., Boudaoud, A., Argoul, F., & Hamant, O. (2011). In vivo analysis of local wall stiffness at the shoot apical meristem in Arabidopsis using atomic force microscopy. *Plant Journal, 67*, 1116–1123.

Milani, P., Mirabet, V., Cellier, C., Rozier, F., Hamant, O., Das, P., & Boudaoud, A. (2014). Matching patterns of gene expression to mechanical stiffness at cell resolution through quantitative tandem epifluorescence and nanoindentation. *Plant Physiology, 165*, 1399–1408.

Nakayama, N., Smith, R. S., Mandel, T., Robinson, S., Kimura, S., Boudaoud, A., & Kuhlemeier, C. (2012). Mechanical regulation of auxin-mediated growth. *Current Biology, 22*, 1468–1476.

Nath, U., Crawford, B. C., Carpenter, R., & Coen, E. (2003). Genetic control of surface curvature. *Science, 299*, 1404–1407.

Peaucelle, A. (2008). AFM-based mapping of the elastic properties of cell walls: at tissue, cellular, and subcellular resolutions. *The Journal of Visualized Experiments, 89*, e51317.

Peaucelle, A. (2015). AFM-based mapping of the elastic properties of cell walls: at tissue, cellular, and subcellular resolutions. *Journal of Visualized Experiments, 89*. http://dx.doi.org/10.3791/51317.

Peaucelle, A., Braybrook, S. A., Le Guillou, L., Bron, E., Kuhlemeier, C., & Höfte, H. (2011). Pectin-induced changes in cell wall mechanics underlie organ initiation in Arabidopsis. *Current Biology, 21*, 1720–1726.

Proseus, T. E., & Boyer, J. S. (2008). Calcium pectate chemistry causes growth to be stored in Chara corallina: a test of the pectate cycle. *Plant Cell Environment, 31*, 1147–1155.

Reddy, G., Heisler, M., Ehrhardt, D., & Meyerowitz, E. (2004). Real-time lineage analysis reveals oriented cell divisions associated with morphogenesis at the shoot apex of *Arabidopsis thaliana. Development, 131*, 4225–4237.

Reinhardt, D., Mandele, T., & Kuhlenmeier, C. (2000). Auxin regulates the initiation and radial position of plant lateral organs. *The Plant Cell, 12*, 507–518.

RoutierKierzkowska, A. L., Weber, A., Kochova, P., Felekis, D., Nelson, B. J., Kuhlemeier, C., & Smith, R. S. (2012). Cellular force microscopy for in vivo measurements of plant tissue mechanics. *Plant Physiology, 158*(4), 1514–1522.

Sampathkumar, A., Krupinski, P., Wightman, R., Milani, P., Berquand, A., Boudaoud, A., … Meyerowitz, E. M. (2014). Subcellular and supracellular mechanical stress prescribes cytoskeleton behavior in Arabidopsis cotyledon pavement cells. *Elife, 16*, e01967.

Sassi, M., Ali, O., Boudon, F., Cloarec, G., Abad, U., Cellier, C., … Traas, J. (2014). An auxin-mediated shift toward growth isotropy promotes organ formation at the shoot meristem in Arabidopsis. *Current Biology, 24*, 2335–2342.

Tirnauera, J. S., & Biererb, B. E. (2000). EB1 proteins regulate microtubule dynamics, cell polarity, and chromosome stability. *Journal of Cell Biology, 149*, 761–766.

Tsugawa, S., Hervieux, N., Hamant, O., Boudaoud, A., Smith, R. S., Li, C., & Komatsuzaki, T. (2016). Extracting subcellular fibrillar alignment with error estimation: application to microtubules. *Biophysical Journal, 110*, 1836–1844.

Uyttewaal, M., Burian, A., Alim, K., Landrein, B., Borowska-Wykret, D., Dedieu, A., … Hamant, O. (2012). Mechanical stress acts via katanin to amplify differences in growth rate between adjacent cells in Arabidopsis. *Cell, 149*, 439–451.

Wolf, S., et al. (2012). Growth control and cell wall signaling in plants. *Annual Review of Plant Biology, 63*, 381–407.

Žársky, V., & Cvrčková, F. (2014). Plant cell morphogenesis, methods and protocols. In *Springer protocols*. Humana Press.

FSC
www.fsc.org
MIX
Papier | Fördert
gute Waldnutzung
FSC® C083411

Kolibri 360 GmbH
Ferdinand-Jühlke-Straße 7
99095 Erfurt, Deutschland
produktsicherheit@kolibri360.de